Dictionary of Inventions

The Wordsworth

Dictionary of Inventions

—

Gerald Messadié

First published in France as *Les grandes inventions de l'humanité.*
Published as *Great Inventions through History* by
W&R Chambers Ltd, Edinburgh.

This edition published 1995 by Wordsworth Editions Ltd,
Cumberland House, Crib Street, Ware, Hertfordshire SG12 9ET.

ISBN 1-85326-357-5

Printed and bound in Denmark by Nørhaven.

The paper in this book is produced from pure wood
pulp, without the use of chlorine or any other substance
harmful to the environment. The energy used in its
production consists almost entirely of hydroelectricity
and heat generated from waste materials, thereby
conserving fossil fuels and contributing little to the
greenhouse effect.

Acknowledgements

Translated from the French by Melanie Hanbury

Adapted for the English edition by Mary Baxter
Min Lee
Erica Schwarz

Chambers Compact Reference Series Editor Min Lee

Illustration credits

Page	
ix	© Boyer-Viollet
21	top: © Viollet/Archives Photeb
21	bottom: © Bibliothèque nationale/ Archives Photeb
30	© Archives Photeb
52	© Historic telecommunications collection, C.N.E.T./Archives Photeb
53	© Boyer-Viollet/Archives Photeb
67	© Bibliothèque nationale/Archives Photeb
77	© Explorer-Archives
85	Drawing by Jacques Cuisin
87	© The Needham Research Institute
94	In *Les mécaniciens grecs. La naissance de la technologie* by B. Gilles. Editions du Seuil, 1980
98	Drawing by Jacques Cuisin
101	In *La dioptre d'Héron d'Alexandrie* 1862; reproduced in *Les mécaniciens grecs. La naissance de la technologie* by B. Gilles. Editions du Seuil, 1980
102	top: In *La dioptre D'Héron d'Alexandrie* 1862; reproduced in *Les Mécaniciens grecs. La naissance de la technologie* by B. Gilles. Editions du Seuil, 1980
102	bottom: In C. Frémont *La Vis* 1928; reproduced in *Les mécaniciens grecs. La naissance de la technologie* by B. Gilles. Editions de Seuil, 1980
104	© Photeb
105	© Bibliothèque nationale/Photeb
108	© Viollet/Photeb Collection
124	© Bibliothèque nationale/Archives Photeb
139	© IBM/Archives Photeb
145	© Viollet/Photeb Collection
151	© Bibliothèque nationale/Archives Photeb
154	© Explorer-Archives
167	Drawing by Jacques Cuisin
173	In *Antique Histoire de qeulques inventions modernes* by Jean de Kerdeland. Editions France-Empire
195	© IBM/Archives Photeb
199	top: Science Museum, London Photograph: La Vie du Rail/Archives Photeb
199	bottom: © Harlingue-Viollet/Photeb
217	In *Antique histoire de quelques inventions modernes* by Jean de Kerdeland. Editions France-Empire
218	© The Needham Research Institute
219	In *Antique histoire de quelques inventions modernes* by Jean de Kerdeland. Editions France-Empire
221	By permission of the Syndics of Cambridge University Library
222	© Royal Asiatic Society
228	© Archives Photeb

Contents

Introduction

Invention characterizes the living being; it attests to an effort of adaptation to the environment. Therefore it is found in the animal kingdom as well as in humankind. In Africa monkeys have been seen to use a stick to get at food which is out of reach: they have invented a tool. This ability is not restricted to mammals: birds can be watched dropping thick shells on rocks in order to break them and then eating the molluscs inside. Perhaps the Californian mosquito shows proof of invention too, as it increases by 200 times its production of the gene which enables it to synthesize the enzyme antagonistic to an insecticide.

Animal inventions are limited, however, and traditionally the history of inventions starts with the emergence of *Homo sapiens sapiens*. Hardly had he appeared but *H. sapiens sapiens* not only adapted to the environment but also adapted the environment to himself. The spears with fire-hardened points served first of all as hunting weapons and then for building fences. Some eight thousand years BC human populations were no longer content with hunting, fishing and gathering. They captured wild animals which they domesticated: horses for draught and for riding, dogs for protection, sheep for meat and poultry for eggs and meat; they also selected crops to cultivate, and therefore founded settled communities. So agriculture began. The countryside changed drastically, through the burning of the woodlands and then by repeated cultivation. Man invented pottery, fixed shaped flints on to shafts of wood, and made different tools according to his needs, such as the scraper, axe, pruning knife and adze. At approximately the same time the settlers in Anatolia and on the banks of the Danube invented a rational organization of their shelters which were to become permanent. The food stores were wisely situated in the middle of the communities. Such was the origin of town planning.

During the Copper, Bronze and finally Iron ages, in the 5th, 3rd and 1st millenia BC respectively, the techniques and consequently the inventions began to diversify. One or more unknown persons invented successively the technique for working iron, then iron with carbon to make steel, then how to make needles, then metal wheels, armour, and cooking utensils which could withstand high temperatures. The historical age, very roughly taken as starting in the 5th century BC, took over from the prehistoric ages.

Then great waves of invasion and commercial trading played a role in the spreading of inventions, comparable to that of swarms of insects in the pollination of plants. The conqueror and the tradesman placed instruments and products from one end to the other of the ancient continents, Africa and Eurasia, until the conquest of the Americas. Despite the work of historians and documentalists, it is impossible to compile a complete history of inventions dating from ancient times to the end of the 19th century, for three reasons.

The first is that obviously no large-scale control of production existed until well into the 19th century; the work of each inventor was carried out to fill his own needs and wishes and so did not resemble that of anyone else. Taking the case of machines for lifting or moving heavy objects, of which we have only a few descriptions, special types of equipment are thought to have been made for different purposes: for raising blocks of stone, lifting columns, unloading boats, lifting these same boats, etc. Hero of Alexandria, for example, invented a system of gears, a technological wonder of the time, then another with an endless screw and a moving bar, and yet another with an endless screw and a cogwheel, each one constituting a proper invention. Hundreds or even thousands of types of lifting instruments have existed throughout the centuries, and one just gets lost in speculation as to the one which was used in the building of the fabulous Colossus of Rhodes in the 3rd century BC, which was 35 m high.

The second reason is that a distinction was, and is made between inventions which were designed on paper and never realized and those which have been put to actual use. In the 15th century the Italian Francesco di Giorgio Martini invented a vehicle on wheels which moved without being pulled by an animal, which was an innovation unheard of at the time. A set of four hand-operated capstans activated reduction cogwheels which made the wheels turn. This extravagant machine, the distant ancestor of the motor car, does not seem to have been built; its originality lies in offering an example of the first vehicle which was not directly powered by animal traction. It is worthy of mention, so it is mentioned here, but the same principle of selection cannot be applied to all unrealized inventions, or a huge volume would be required. The breathtaking invention of a certain Carron, an engineer from Grenoble, is not included. In 1891 he had the idea of dropping passengers enclosed in a metal capsule from the top of the Eiffel tower! The aim of the invention was to produce strong emotions in the passengers. Carron suggested making a 60 m-deep pond at the foot of the tower to receive his shell, but it is not known exactly how the clever Carron envisaged the recovery of his capsule. The floor was equipped with powerful springs to soften the impact, and the arch of the capsule was lit — with a gas light? — to prevent the passengers being frightened of the dark. The price to be charged for producing the 'strong emotion' was twenty francs.

The third reason that an exhaustive inventory is impossible is that many inventions are undoubtedly lost for ever and others have been lost and found again, or reinvented. For example, not all the works of the inventors of the famous Alexandrian School, to mention just one extraordinarily fruitful centre of technological activity in the ancient world, are known. Before printing, the only documents describing inventions were manuscripts, of which often only a very few copies existed. Nothing is more vulnerable than a library. Fire and occupying armies are its worst enemies — in only a few hours irreplaceable knowledge can disappear in clouds of smoke.

Inventions can also disappear as a result of civilization, or of regression. In this way during the Middle Ages certain principles were 'forgotten', such as triannual systematic rotation and soil fertilization which the Romans had carried to a point of perfection remarkable in its time.

Two examples can be given of such eclipses. The first is that of the steam engine, which was most certainly invented around the 1st or 2nd century BC by Hero of Alexandria. This invention was submerged until the 17th century. All the necessary elements for the rudimentary but nonetheless revolutionary manufacture of a steam machine were available in Hero's time: the boiler, possible as a result of the manufacture of hard steels and the invention of soldering; the piston, specimens of which were made which seem to have had a satisfactory bore; the transformation of a rectilinear movement into a circular movement which would also lift, as in the Archimedean screw. Did the steam machine fail to see the light of day because slavery made recourse to mechanical energy pointless? Was there a scarcity of fuel? There was, but this is not sufficient to explain the absence of an invention which was later so successful. This is a riddle without an answer.

The second example is that of the established use of maps. From the 3rd century BC, Greek mathematicians made good use of the lessons of Pythagoras, and plane trigonometry could be applied to geography. At the same time, Eratosthenes had calculated the radius of the Earth to be about 6000 km. Precise geographical maps were certainly possible, at least for the parts of the world which were known (the Greeks also made astronomical clocks which enabled navigators to calculate the distances which they had travelled by sea with great precision). In theory, antique maps of the world could have been made using trigonometry. When Christopher Columbus, however, enquired about maps of the world, for example from Toscanelli and Behaïm, he could only obtain maps which had considerable errors: the Earth's circumference reduced by a quarter and the breadth of Eurasia shown to be 180° instead of 126°. As a result Columbus was not only unaware of the presence of America, but he also seriously underestimated the length of the crossing, which caused a mutiny among the sailors on his expedition. How did the Greek cartographic knowledge come to be lost? Again we do not know.

The case of Chinese inventions is particularly astonishing. Their number and their advancement is impressive: suspension bridge, hydraulic piston motor, chain transmission, magic lantern, drills, mechanical clock, sluice, printing, flame-thrower, and the cardan or universal joint (which for centuries was only used to stop perfume burners from tipping over!). Most of these were invented during the first few centuries AD, some before that. It is baffling that so much enrichment could have fallen into oblivion. Also, it shows that technology alone does not ensure political supremacy if it is not accompanied by trade activity.

At the time when Chinese technology shone with a brilliance comparable only to that of the Alexandrian School, plain technology did not sell. The Chinese hardly considered exporting drills, for example, which would have been very popular in other countries, and it was not until the 17th century that the Englishman Robert Hooke reinvented that wonderful mechanism, the universal joint. There has been a dreadful wastage of inventions over the years. Apart from the absence of commercial trading, the obsession with the Devil must undoubtedly be held responsible in Europe. He is willingly forgotten these days, but any invention which was a little too clever, a bit too surprising, was likely to seem like the fruit of a pact with Satan, and for a long time there was a danger that the inventor himself would end up on the stake. It is hardly possible to think of the Middle Ages in Europe as anything other than a deplorable age of darkness and fog. Even the major innovation in the technology of building, the gothic arch — and the revolutionary alternative to the supporting walls, which allowed the master glass-workers to display their craft in large windows — was for a long time suspected of 'barbarism'!

Until the end of the 19th century many inventions were lost in this way, often several times successively, such as the compressed air rocket which already existed at the time of Henry IV (and even in ancient times) and which has been rediscovered periodically, or the typewriter, one rudimentary model of which had already been made during the reign of Louis XIV, or the lift (one perfectly reliable handle-operated example of which was installed at Versailles under Louis XV).

Like a moth which flutters at the lighted window, the human mind caresses numerous 'modern' inventions throughout the centuries, leaves them half observed, sometimes half invented, and then returns to them according to circumstances and requirements. The method of air cooling which the Hindus had engaged from ancient times — putting wet mats in front of the doorways of buildings — was apparently forgotten throughout the whole of the Middle Ages. Leonardo da Vinci invented a hydraulic system for cooling Isabella d'Este's home but this was not followed up. The first European building to be mechanically cooled was the British Parliament in 1837, three years after the Frenchman Peltier had discovered the thermic inversion principle or Peltier effect, which is the basis of our air-conditioning systems and which was not to be fully used until a century later, in the United States.

These unwitting repetitions are a result of lack of documentation. For a long time an inventor would not know of a similar invention by a predecessor. For example in France, the exclusive rights of an invention were protected only by the letter of patent which was granted by royal authority. This letter was archived only in Paris and so a provincial inventor would not know about it except by taking a stagecoach or a boat and going to Paris. Probably the catalogue of letters of patent was not well categorized either, which would not have made research any easier. The actual invention patent did not appear until the revolutionary law of 1791, but the archives were to be transferred from one minister to another, notably the Home Office and Agriculture, for a long time. The inventions only spread as a result of the fame of the inventor, or thanks to his patron. Often their fate was uncertain.

This situation did not change until the beginning of the 19th century, when trade and industry at last began to take inventions seriously, and legal proceedings might be taken against plagiarists. No one was ignorant any longer of the fact that a fortune could be made from a patent (although Sainte-Claire Deville, who invented a process for separating aluminium in 1854, parted with this patent at a very low price).

Since the 19th century there has been less

risk of a patent being 'lost', but this does not mean that it is always recognized at its rightful value. Napoleon shrugged his shoulders in front of the steam barge made by the American Fulton at its first (admittedly imperfect) demonstration on the Seine, saying that it had no future. Thus also Edison claimed the radio had no future, and the Wright brothers instituted proceedings against Clement Ader, whom they did not know about, or pretended not to know about, for inventing the aeroplane before they did.

One invention above all has had a decisive influence on the destiny of other inventions, that of printing. By enabling the fixing of both theoretical and technical knowledge on paper in several copies, Gutenberg finally made it possible to save inventions from destruction, or prolonged disappearance. No one knows how many inventions perished in the three successive destructions of the library in Alexandria, the largest store of knowledge of the ancient world. Even though the scientific press began quite late (the two first scientific newspapers were the *Erbauliche Monaths-Unterredungen* ('Monthly edifying discussions'), first produced in 1663 in Germany, and the *Journal des Sçavans* ('Scientists' newspaper') created by Denis de Sallo in 1665 in France), the simple fact of being able to print scientific reflections from the 15th and 16th centuries guaranteed a certain conservation of technological knowledge. The widely circulated magazines and the bulletins of scientific societies in the 18th and especially the 19th centuries were to renew, maintain and enrich technological knowledge in an exponential progression, helped by the growing literacy of the masses.

As well as printing, two other important factors gave a vigorous boost to the course of inventions, at least in Europe. The first was the intellectual liberation following the Reformation, which explains why for nearly two centuries the Protestants took pride of place in this domain. Once the scientific quietism jealously maintained by the Vatican (to which the cases of Copernicus and Galileo testify so eloquently) had been swept aside, discoveries and inventions arising out of scientific curiosity proliferated. Not that Rome had specifically curbed the spirit of inventiveness, but the impetus to explore the phenomena and laws of nature had been considerably stimulated by the intellectual freedom born of the Reformation.

The second factor to spur the inventive spirit was the craving for energy. Since Archimedes, man had endeavoured to channel the natural forces in order to make use of them. But the two 'Dark Ages', that of the Barbarians, followed by the Middle Ages, considerably slowed down the search for energy. This was not taken up again until about the 17th century, reaching its full expansion in the 19th century. Industry, the nations' peaceful weapon, could no longer be satisfied solely by human energy, and after having almost exhausted the mechanical techniques and 'man-power', it eventually set out to work the virgin ground of steam and electricity. It should be noted that countries with a settled government were infinitely more favourable to inventions than nations which had no judicial structure.

The above observation explains why the Latin countries remained absent from the panorama of inventions for a long time, and why the huge majority of inventions was consequently concentrated in the countries which had reached the most advanced level of social and civil organization, the European countries. It is not surprising therefore that the technical conquest of the world was closely followed by its political conquest. Until World War I the great colonial powers were also the great technical powers: Great Britain, the Netherlands, Germany and France. During the 20th century, the inventive spirit ended up anointing some lands as technologically impoverished as, for example, certain countries in Asia.

Inventions therefore have a history which is closely linked to history, which is why we have divided invention into two distinct periods: one, which is the longer, where the inventiveness is still a privilege, dating from ancient times to the mid-19th century, and the other where genius proliferates over the whole planet, which we have placed — arbitrarily — at the end of the 19th century.

Inventions do not arise only from commerce and the desire for power. Important as these two factors are in the development of inventions, they do not constitute the essential driving force. The seagull which carries a clam in its beak, places it on a wall, then goes to pick up the biggest stone it can manage and drops it from a height in order to break the shell has invented a technique which is inspired by neither the spirit of commerce nor the desire for power. This technique enables it to obtain its food reasonably quickly and in return for a little ingenuity; it prevents the bird from having to test the strength of its beak against the mollusc and being obliged to go and find something else to eat, which could take time. Thus it saves time and consequently energy: this is the goal of absolutely all the inventions which have been made since humanity began. The first suspension bridge thrown by the Chinese over a gorge in the Himalayas some two thousand years ago avoided the considerable detours which were costly in time and energy; similarly the first geographical map was to prevent wandering which could be fatal. Printing banished the need to copy a text several times by hand, and weighing scales meant that the exact weight of a commodity could be known immediately. The motor car, like the aeroplane and the paddle steamer, is a means of travelling more quickly, and therefore further, with the least possible effort. All inventions are included in this absolute rule. If some have seemed to develop with excessive slowness, it is because they were not immediately recognized as a means of conserving energy. If some seemed at first to be of little interest, it was because they consisted of a first, blind step towards a vaguely perceived goal. Thus the first electric light bulbs were considered to be a curiosity without much of a future.

It is easy to forget that the human brain obeys the same fundamental law as any other energy-carrying system. Just like the electron, the bacterium, the seagull and the elephant, it has a tendency to descend to its lowest energy level. Invention is the daughter of a divine laziness, and the whole of technology, even the most brilliant, illustrates this law which is known under the name of the law of entropy. It is much more convenient, though hardly more stimulating intellectually, to have illumination by turning on a switch than by lighting a firebrand in the fire, trimming the wick of a lamp and then lighting it (not to mention having to change the oil in the lamp). How much easier it is to open a tin of pâté than to shoot, pluck, draw and cook a gamebird or another type of fowl!

Invention appears to be a paradoxical expenditure of energy meant to lead to a reduced energy expenditure. It is comparable to the modern holidaymaker who undergoes a thousand chores in getting ready for the holiday, in order then to give way to idleness. But even then the 'reward' must be proportional to the effort, as it was in the case of the crucial invention of matches.

This book also aims to illustrate the often immense gap which separates the imagination from the viable invention. One famous example of this gap, and one which is widely misunderstood, is the engineering work of Leonardo da Vinci. Leonardo was overflowing with imagination, as attested by his sketches, but very few of his inventions withstand examination. The blockhouse mounted on wheels and decorated with an impressive battery of cannons, in which so many have seen the ancestor of the tank, is an unrealizable fantasy. Its weight almost rules out its effective use in the country, and one wonders how many horses would have been required to move it. It is impossible to imagine twelve horses and about fifty cannon loaders inside this stifling contraption. Equally unrealizable is his device of articulated wings which were to enable man to fly; the experiments carried out by Lilienthal at the end of the 19th century proved this fact. Leonardo's device was inspired by the flight of birds which he had watched attentively, but he underestimated the problem of the aerofoil and manageability. He certainly cannot be reproached for having dreamed, as did so many other engineers at the time; however, it is clear that he was first of all, like most of the artists of his time, an inventive artist and a technician — but a technician without authentic technology. Hence the proportion of tin (6%) which he advocated for the casting of a bronze cannon was wrong (he would have needed 12%; nevertheless it was his proportion that the casters applied empirically).

Theory and practice eventually began to unite in the 19th century. Until then the theoreticians displayed an immoderate faith in their own reasoning. This is perhaps not surprising, as science has a tendency to become a substitute for religion; it is an expression of philosophy, a way of looking at the order of things. Also the overall approach of scientists, technicians and inventors is to explain nature rather than to observe it. Thus the great Euler made an impressive blunder which caused a delay, or more precisely a halt, in the progress of ballistics. Artillery-men of the 18th century endeavoured to improve the accuracy of their cannon-firing with little success. An Englishman, Robins, who does not enjoy the same prestige as Euler, noted that one of the causes of imprecision was the deflection of the bullet's path due to friction against the bore of the gun; he suggested remedying this by scoring the bore longitudinally. But Euler, sure of his own calculations, rejected both Robins's observations and his solution. It was more than a century before they were recognized as fully justified. There are innumerable examples of this type.

This is why real inventions are quite rare. Until the mid-19th century, when technicians decided to consign their philosophy to the top shelf, tempering their superstition with numbers and mathematics and at last accepting the necessity of experimentation, invention was imbued with an almost fortuitous character. This is particularly noticeable in chemistry and physics, the two domains where ideas are the most cumbersome and where observation is the most necessary. It is noticeable also in medicine, where proper observation of the human body did not begin until very late. Before that completely erroneous ideas were lovingly entertained, such as the nutritional origin of diabetes and gout. The sectors where there is most inventive development are warfare and armament, industry and mechanics, and the exploitation of energy and transport.

As previously said, no list of inventions could be exhaustive: it is inevitably selective. This selection has been made by elimination. Thus the book does not include minor inventions, such as the machine for cutting goose feathers; eccentric inventions, such as Carron's metal capsule; and several of the inventions which are now obsolete, such as the innumerable seats, catapults and battering rams used in ancient military warfare. The list is further limited to inventions which have persisted successfully and have affected the evolution of society and of our modern daily life. Particular attention has been paid to chains of inventions, showing the process of trial and error involved in the development of inventions by humankind.

The selection is an arduous task, for even after its demise a former invention might have had a profound effect on its era and also have generated modern-day inventions. Such was the case with the sundial and the clepsydra, the ancestors of our clocks and watches. Significant inventions such as these have been retained, on the grounds that numerous inventions are often profoundly modified or remade years later. The inclusion of the calculating machine in the inventions is evidence of pure semiotics, for there is no longer anything in common between Pascal's adding machine and our calculators. Moreover, at which point should the distinction be made between the adding machine proper and the mathematical machine? The one developed from the other: it was for doing calculations that Lord Kelvin invented the first mathematical machine.

When writing a history of inventions, however succinct, it is important that derivations are taken into account. Thus, underlying the clock and the adding machine there is the elementary invention of the cogwheel, just as at the base of Vaucanson's weaving loom and Babbage's machine there is the perforated card. These derivations are obvious. When one tackles the Greek *baroulkos*, however, the derivation is purely thematic, for it seems that the Chinese, who made the first geared astronomical clocks, had no knowledge of it, any more than the first European clockmakers did. Thus the 'derivatives' are described in the same way as distinct entities in the text and indicate the origins of the invention.

Presented in this way the question of derivatives seems deceptively simple. In fact, compiling a catalogue of the great inventions, especially spanning ancient times to the 19th century, when science and techniques were much less differentiated than they are today, presents difficulties similar to those faced by the taxonomists who endeavour to establish the evolution of living species. Thus the steam turbine and the internal combustion engine, both distinct technological entities, are both derived from the first steam machine, Hero of Alexandria's aeolipile. That the turbine is derived from the aeolipile is obvious enough, but the combustion engine? This is derived from the aeolipile through the steam engine, but it could not have come about without Ctesibius's invention of the suction and force pump, which introduced the principle of the cylinder and the piston. Problems of this kind have been many in the realization of this work, especially in the domains of energy, mechanics and industry. It is obvious, for example, that the propeller originates in the windmill, but the beginnings of it can also be found in Archimedes' screw. Did it become the propeller, a totally specific technological entity, only from the moment it was used for aerial or aquatic propulsion? Or should this invention be thought to date from the time it was first used? In defining the derivatives according to the principle outlined above, the propeller is presented as a distinct entity and is placed in the chapter on transport, whilst the windmill is to be found in the chapter on energy and mechanics. Arbitrary classification it may be, but it is inevitable if the reader is to be offered a convenient approach to the great inventions.

Every theory is a potential invention, but so many theories would have had to be included that this book would have tripled in size. Even so a major philosophical problem would have arisen: how to differentiate a law, which is at once a verified theory and a discovery, from a theory which has not yet been proved, but which will be. Thus in the 3rd century BC Eratosthenes deduced on the basis of mathematical calculations that the Earth is round. Since he could not provide proof, his theory was disregarded in favour of the old idea that the Earth is flat and round like a plate, comprising only Europe, Asia and North Africa. The sphericity of the Earth was not to be established until several centuries later. It would have been necessary to include Eratosthenes's theory (or rather to reconstitute it) since it was correct. Here the preference has been to rule out a great number of theories on the grounds of Karl Popper's postulate, which states that any theory is correct until it is proved that it is false. And we will go further: that it is false until it has been completed. Since one could maintain, metaphysically, that no theory is ever complete, it seemed wise to keep to those which have founded modern-day disciplines, such as linguistics.

On the other hand, cultural inventions which seem to have significantly altered our life style have been included. An example of this is the serialized novel, without which the cinema, among other arts, would not be what it is today. Without Dickens and his serialized novels, we would have had neither Hitchcock, nor adventure films; perhaps we would not have had the likes of Dostoyevsky or Maupassant, to mention but a few. It was with the serial that the novel became differentiated from the narrated story, and, most importantly, gave rise to the 'popular' novel.

Some disciplines which do not seem to pertain to the domain of inventions, such as the above-mentioned linguistics, are also included. At first linguistics was a barely differentiated derivative of grammar, which is undoubtedly not an invention in the same way as the electric light bulb. It nevertheless plays a crucial role at the end of the 20th century, in the development of computers capable of translating one language into another and recreating human speech. By the end of the century computers will speak and, further, will obey vocal commands. The analysis of grammatical structures and their equivalents in other languages, one of the objects of linguistics, plays a major role in these experiments.

There is also psychology, apparently a philosophical discipline, but one which during the first years of the 20th century was to lead to the establishment of the intelligence tests which subsequently became so widely used. These have become so commonplace that it is difficult to imagine them as an invention. However, there was a time after which psychology lost the most obvious part of its intuitive and empirical character in order to offer a technique for quantifying the apparently unquantifiable, that is, the capacity to apprehend the elements of a situation. This was the moment of invention.

In including these disciplines and some

others in this work, the aim has been to enrich the general notion of invention, and also to demonstrate that Saussure merits the title of inventor no less than Edison, and that intellectual tools as precious as a milling machine or matches can equally well exist.

There is one final problem to be tackled, both to ease our consciences and to forestall critics: should this work deal only with inventions which have been realized, (which might give it an excessively technological slant), or should it also mention those which have only been conceived or half realized? After all, why omit the aerostat which was to reach the Moon by the effect of the evaporation of the dew, as described by the genuine Cyrano de Bergerac? After all, it is impossible to describe the history of the adding machine without taking account of Babbage's vain attempts to make a universal digital adding machine. The true significance of these attempts came to light only in the 20th century: Babbage's genius, as well as the obtuse avarice of the Prime Minister Sir Robert Peel, is now clear to those interested in the genesis of computing. Therefore inventions of which the realization has been blatantly or almost certainly prevented by misfortune alone will be included.

It is often proposed that nine-tenths of known inventions date from the 20th century. While this is an exaggeration, it is true that a great number of inventions have been made in the 20th century in comparison with all of the preceding centuries put together. To unite all the important inventions in a single work would be impossible, without creating a very large book. This is why it has been decided to divide the material into two parts, of which this book constitutes the first. All inventions up to around 1850 will be found here. This is just a convenient cut-off point, a date in the industrial age after which it seems that humankind has become resolutely 'modern'. From this time inventions have begun to be considered not as a curiosity of science but as a small step closer to the economy of human energy. A few inventions will be found to 'spill over' into the second half of the 19th century: the inventions whose history it seemed specious to split up according to an arbitrary date division.

These pages do not aspire to constitute a new history of technology. Some excellent works already exist which review the schools and movements and the inventive human spirit which have influenced society. We have of course endeavoured not to bore the reader, but it is also true that we have not aimed to offer a work for sheer entertainment. Rather these pages are considered as offering a contribution to a new history of technology following a trend which has developed only in the second half of the 20th century. A certain arrogance within contemporary industrial society has let it be believed that technology was finally born with industry and that apart from using the wheel, the ploughshare and the nail, the preceding centuries tinkered about and floundered in ignorance. It is wrong however to claim that technological intelligence was born with the Industrial Revolution; brains of a similar capacity to those of Edison, Hooke and Babbage did exist before.

Archaeological excavations continue to reveal the technical abilities of our ancestors, and remind us that many inventions originated before our own technological age.

Agriculture and Food

The need for regular food caused *homo sapiens* to turn from hunting, fishing and gathering to the invention of agriculture and the rearing of livestock. It was a very slow invention which lasted for thousands of years, and for which no one can claim paternity. The habit of harvesting which was acquired between the 7th and 6th centuries BC was the very basis of civilization. Indeed, it enforced settlement, which gave rise to towns, then states, politics, technology ... and wars. The cultivation of the ground led to that of the mind. Being isolated geographically, and therefore also genetically, the peoples began to differentiate and their religions to diversify.

One change has come about since the first grain harvest: the intervention of industry. Agricultural mechanization has greatly reduced the agrarian class, causing a radical modification of society. Food preservation and then the whole food industry have certainly eliminated the threat of food shortage in the industrial countries, which persisted in Europe until the 18th century and later in some countries. Now replete, the industrial societies have adopted dietary rules.

Though of small significance, the diversification of cultures, helped by voyages of exploration, and the practices of agricultural selection have also caused integral changes in plant and animal species. There is no longer much in common between the Mexican corn of yesteryear and modern corn, and our farmers really would be surprised if they saw a Middle Ages pig, a remarkable creature with flat flanks and powerful bone structure. The arrival of the potato has changed the diet of half the world and the horse is now used virtually only for competitions and races.

Biological battle against insects

Anon, China, c.3rd century

The biological fight against harmful insects has been practised systematically in the West since the 1950s after the failure of **industrial pesticides** at this time, and was known in China at least at the end of the 3rd century, as attested by a text dating from 304 on the agriculture of the regions of the south of Hsi Han (Xi Han). This text mentions the use of a variety of 'carnivorous' ant to destroy the insects which were attacking the mandarin tree orchards. Bamboo pipes leading from one tree to the other enabled the insects to permeate the whole orchard. This method of pest control subsequently spread from China to the whole of citrus fruit cultivation. The English used it in the 18th century by introducing **menates** into Mauritius to conquer the red grasshoppers which were devastating the sugar-cane plantations. In 1888 the Americans saved their citrus fruit plantations which were threatened by the parasite *Icerya purchasi* by introducing an Australian beetle, *Rodolia cardinalis*. Around 1960, however, the use of industrial pesticides again took priority over the biological fight before it could be restored. The 1980s and early 1990s have seen a renewed interest in biological control, with the trend towards organic farming.

Canned food

Anon, 4th or 3rd millenium BC; anon, China, 1st millenium BC; Appert, c.1780; Pasteur, 1868–1870

Anxiety over the preservation of food is undoubtedly as old as humanity. Ever since man hunted, he must have realized that the meat of his game would go rotten and become harmful in certain circumstances, and that in others it could be dried and eaten quite some time after being killed, even though it lost some of its organoleptic qualities. In any case it is known that from the 3rd or 4th millenium before Christ, and maybe even before that, man used certain caves for storing his game which then dehydrated and dried without going rotten. The Chinese seem to have mastered three techniques of preservation by the 1st millenium: the first, by **salting**, the second by using **spices**, and the third by **fermentation in wine**. The three techniques were equally judicious: the salt created an environment with a pH which slowed down bacterial fermentation; the spices also exerted an antiseptic action, though a less powerful one; and the fermentation in wine prevented microbial fermentation through the bactericidal action of the alcohol. The two latter techniques were the most often associated. Moreover, the Chinese had improved the process of **drying** with that of **smoking**. It is with the Chinese therefore that the actual invention of the preservation of food began.

These techniques hardly varied for at least two and a half millenia. There was an obvious problem with all of them, which was firstly gastronomic and then dietary: these foods always tasted strongly of salt or vinegar. Apart from the making of

> One can easily see why the appertized foods aroused much interest among the maritime authorities, since they allowed some variety to be introduced into the crews' food as well as being much more convenient to store. The prefect of the maritime region of Brest was one of the first people to taste some which had had three months' preservation and he thought they were excellent; they were garden peas.

conserves, or meat cooked and then covered in fat, which seems to have first appeared in France in the Middle Ages, there was no progress in the domain of the preservation of food until the invention of **appertization**.

Nicolas Appert, a confectioner in Paris, was the first to have the idea, in 1780, that applying heat to sealed jars could prevent the fermentation of food; he began to put it to the test. The idea was so remarkable that it was only Pasteur, nearly one century later, who proved that heat actually killed the bacteria responsible for fermentation. In 1790, Appert verified that his idea was indeed correct, and he founded a business to sell meat, milk, vegetables and fruit all preserved in sealed bottles. In 1810, once the Frenchman Pierre Durand had patented his receptacles which were coated on the inside with tin, the preserved food industry began to spread. In 1812 the Englishmen Donkin and Hall purchased both the Durand and the Appert patents and produced the first preserves in metal boxes, an example which Appert himself followed. From 1815 canning industries were created in numerous countries. Appert subsequently extended his invention to the preservation of wine.

The discovery of the role of bacteria in fermentation was to lead Pasteur to design the process of **pasteurization** between 1868 and 1870, which was meant for the limited preservation of food. This consisted of heating the foods to a temperature of about 70°C for a short time, then quickly cooling them. Unlike appertization, pasteurization retains the maximum amount of flavour in the foods, since it does not involve cooking. However, it does not allow long-lasting preservation.

Cereal crops

Anon, Mesopotamia, c.9000 BC

It is estimated to have been after the final glaciation, which occurred 15 000 years BC, that the lifestyle in the ancient world, that is, in Eurasia, changed from a **nomadic** one of **hunting and gathering** food for survival to one of **settlement** and **agriculture**. This fundamental change, which could be explained by an amelioration of the climate and a reinforcement of vegetation is at the basis of the formation of communities and a rationalization of eating habits. Indeed, agriculture enabled the change from the threat of almost chronic food shortage to that of accidental food shortage; it also effected the change from an almost exclusively carnivorous diet to one that was more varied, and therefore more healthy. It led eventually to **breeding** (see below, p. 16).

The most ancient trace of cultivation discovered to date is that which the American Robert Braidwood brought to light in Jarmo, in ancient Mesopotamia, now in north-eastern Iraq: it consists of fossilized grains of **wheat** and **barley**. Analysis of the grains revealed that they were indeed cultivated varieties, for they had **peduncles** on their stems, thus differing from the wild parent varieties which had no peduncles.

Eleven thousand years ago, therefore, man had identified some **wild grasses**, gathered the seeds from them, recognized them to be edible and had the then extraordinary idea, for nothing was known about botany and plant growth, of sowing the seeds to obtain new crops. It is possible that this invention proceeded from the observation of the grains at the start of **germination**. In any case this seemed to be

> Corn absorbs all the nutrients from the earth and if this is not left **fallow** before being sown again it becomes impoverished quickly. It undoubtedly took a few centuries before the first cultivators discovered this law, as well as the importance of **crop rotation** and **fertilizer** (see p. 22 and p. 14).

imbued with the character of magic because for thousands of years germination was associated with divine power, in particular with a fertility goddess who later served as the model for the Greek goddess Sibyl. The first two types of wild grasses which were used in controlled cultivation were barley and wheat. It is certain that for centuries the harvest was difficult and disappointing, with the ears bursting spontaneously in order to sow their seeds in the wind, following their natural method of reproduction. It is possible therefore that the first farmers harvested their crops shortly before they were fully ripe. It is more likely, however, that the ears which they harvested, those still holding their seeds, were those of which the **spikelets**, that is, the whole grain with its skin made of two **bracts**, had remained attached to the main spike of the ear by their **rachilla**. Thus they carried out a **natural selection** which from harvest to harvest led to the creation of a species of domestic corn from which the seeds did not burst upon ripening, resulting in much more fruitful harvests.

It was in the New World, now Mexico, that the Indians seem to have developed the cultivation of a crop which served as their basic diet for about seven millennia, that of **maize**. The origin of domestic maize remains vague, however, for it is a cereal which cannot free its seeds without the intervention of man. Specialists are divided on the identity of the plant which gave rise to domestic maize: some even think that it was Asian. The fact remains that the ears of Mexican domestic maize were much smaller than those of the maize we know today; they measured only 2.5 cm in length. It must have been by the practice of **cross pollination** that the Incas of Peru managed to improve the size of the maize, and it is thought that the Mexican maize owed to the Inca plants the new strength that it had from the 8th century BC.

The final two great stages of arable agriculture were the domestication of **millet**, which took place in China 4000 years before Christ, and that of **rice**, which took place about 3200 years before Christ, also in China. It is nevertheless possible that rice had been cultivated previously in South-East Asia, and then introduced later to China.

Chemical pesticides

Anon, China, 7th–2nd century BC; anon, France, 1763

An advanced knowledge of the properties of numerous plants put the Chinese in the position of being able to practise disinfectant **fumigation** of their houses in the 7th century BC and also **gas warfare** in the 4th century BC (see p. 218). Therefore it is quite possible that they made the first known **agricultural pesticide** around the 1st or 2nd century BC, though an even earlier origin is likely. This was **powder of pyrethrum**, an extract from the flower of a plant in the chrysanthemum family. The use of **insecticides** prior to 1763 is uncertain, this being the date when **tobacco** leaves began to be ground up to fight off **aphids**. It is known for certain that in 1800 **copper sulphate** was used more and more in the '**Bordeaux soup**' which was to combat **mildew** and **phylloxera** on the vines, as were **hydrogen cyanide**, **lead arsenide**, **nicotine**, **rotenone**, **oil**, **creosote** and **terebine**. The development of **synthetic pesticides** did not begin until the 1930s.

Only about twenty years later it was recognized that the persistence of the molecules in these substances, their weak selectivity and their accumulation in food chains, the soil and the phreatic layers were causing effects as harmful as the evils that they were meant to combat: the disappearance of useful species and insect resistance.

Condensed milk and powdered milk

Borden, 1856

With a view to facilitating the **transportation** and **storage** of milk, in 1851 the American Gail Borden had the idea of drying it. Depending on the stages of this operation he obtained either condensed milk or powdered milk. When it was patented in 1856 the invention aroused hardly any interest, but later the American Civil War proved its importance and Borden's fortune was assured.

Fertilizer

Anon, Mesopotamia, Egypt, 1st millenium BC; Cato the Elder, 200 BC; Van Helmont, 1609–1644; Glauber, c.1630; Woodward, 1699; Saussure, 1804; Boussingault, 1834; Liebig, 1840

The use of fertilizers in agriculture constitutes one of the most disconcerting chapters in the history of agricultural inventiveness. Mid-way between discovery and invention it disappeared for ten centuries.

The role of fertilizer was almost certainly discovered by observation, about three thousand years ago. The use of **human excrement** and **manure**, mixed with **vegetal waste**, is attested in Mesopotamia in the 1st millennium BC, as it was in Egypt, where the silt of the Nile could also be used. It then spread throughout the Mediterranean basin, and the Romans codified it. Around 200 BC, Cato the Elder mastered the problems of **soil fertility** to a certain extent by recommending the addition of **lime** to acid soils, and manure to all soil, in addition to the cultivation of **leguminous plants** which fixed **atmospheric nitrogen**. Cato's advice is all the more surprising since the fixing of nitrogen by plants was not investigated scientifically until 1834 by the Frenchman Boussingault, and its mechanism established in 1891 by the Russian Winogradsky.

The progressive decline of the use of fertilizer after the fall of the Roman Empire is also very surprising. Equally incredible is the fact that China and India, countries where famine soon took on endemic proportions and where farming techniques were nevertheless often very developed, did not turn to the use of fertilizer until modern times.

The man who rediscovered fertilizers after ten centuries of profound ignorance during which plants were thought to take their nourishment from humus was in fact a 'reinventor', Jean-Baptiste Van Helmont, a Dutch alchemist with often fantastical ideas who carried out a famous experiment. In 1609 he planted a 2.5 kg bush in 100 kg of soil; five years later the bush weighed 84.5 kg and the soil had lost less than a hundred grams. Van Helmont discarded the foundations of **plant physiology** and proved that plants did not feed from the soil but from water and 'alchemical elements'.

The practice of using fertilizer was still not fully accepted in agriculture. Van Helmont was continuing his research — which was later interrupted by his death in 1644 — when around 1630 the German J.R. Glauber, another alchemist working in Amsterdam, postulated that **potassium nitrate** or saltpetre is the principle of life and a choice fertilizer — which is perfectly correct, as he proved it to be. At his death he left a scientific testament in which he explained how to increase Germany's agricultural resources. More attention was paid to Glauber than to Van Helmont and interest in both natural and chemical fertilizers began to arise. Glauber thus occupies the position of predecessor which time could only confirm: he is the true inventor of **chemical fertilizers**.

His direct successor was the Englishman John Woodward, who in 1699 cultivated plants in rainwater, Thames river water, water from the London sewers, and water to which he had added some garden mould (in which there were probably traces of saltpetre). He established that the more 'impure' the water was, the better the plants grew. More than a century passed before the Swiss Theodore de Saussure outlined the necessary proportions of fertilizer for plant growth, but fertilizers still were not widely used in farming. In 1840 the German Justus von Liebig demonstrated the fertilizing action of **phosphorus** treated with acid. Two years later the Englishman John B. Lawes founded the chemical fertilizer industry, following his invention of the process of the production of **superphosphates** by the addition of **sulphuric acid** to **phosphate rocks** (apatite), which was derived from Liebig's observation. It was only at the end of the 19th century that the use of fertilizers spread. Twelve centuries had passed since the time when it had first appeared in agriculture.

Liebig's law has had a great future since 1840. It states that the growth of a plant is limited by the assimilable element which is of the lowest concentration in the environment. It was first applied only in agriculture, but is now used in many ecological applications.

Harvester

Anon, Gaul, 1st century

One of the peculiarities of agricultural technology is that the harvester, which appeared at a very early date in history, fell into disuse, not to be 'rediscovered' until the 19th century. Pliny the Elder, who lived in the 1st century, describes a Celtic cart equipped with a **toothed blade** and pushed by a draught animal. This would have been used in Gaul to reap the ears and the straw. The machine survived for four centuries but apparently only in the Celtic region of Gaul; perhaps it was not to the liking of the farmers. The harvester was reinvented in the first years of the 19th century in the United States, as a dual-purpose machine, such as Lane's **combine harvester** (1818), and the **self-binder** made by John F. Appleby (1858) and by Walter Wood (1871). The first successful self-binder was that of the American Cyrus Hall McCormick, which was launched in 1890 and which reaped and bound the sheaves simultaneously; it proceeded through numerous successive improvements, such as the American Dorsey's **swather**, invented in 1856. Mechanized agriculture resumed with the antique invention after an 'interval' of nineteen centuries.

Incubators

Anon, Egypt, 7th century BC

During and possibly before the 7th century, the Egyptians used to incubate eggs in '**chicken ovens**'. There was such an abundant production of chicks that they were sold in baker's dozens.

The invention of **zoological observation** had evidently ensued, for several reptiles and all birds (and Monotremata, such as the duck-billed platypus, which of course the Egyptians did not know) practise controlled incubation. Some species even use the heat generated by decomposing grass, volcanic soil and solar radiation reflected by the ground.

Irrigation

Anon, Mesopotamia, 9th–8th century BC

One of the most important inventions for agriculture, the distribution of water to cultivated ground via a network of channels, has undergone an astonishing variety of destinies following different civilizations and ages. The oldest known verifiable mention of it dates back to the 9th or 8th century BC in Mesopotamia. Helped by the proximity of the Tigris and the Euphrates, which assured a constant supply of water, but also threatened by the violent flooding of these rivers, the Mesopotamians built networks of channels which were equipped with **regulating sluice-gates**, as a result of which their farmland became particularly fertile. In Egypt too a vast irrigation network, which was fed partly by Lake Moeris, an enormous artificial reservoir, enabled the extension of the cultivated land right up to the desert. Paradoxically, there was no irrigation network in Greece, either during the classical or the hellenistic eras, and the land was reduced to dry cultivation. The Romans on the other hand practised extensive, systematic irrigation, as a result of which they succeeded in transforming the North African provinces into the 'foodstores of the Empire'.

Livestock rearing

Anon, between 8500, Mesopotamia and 1000, Russia, BC

The rearing of livestock is one of humankind's three greatest inventions, the others being agriculture and metallurgy. Agriculture put an end to the hazards of gathering food and reduced the risk of food shortages; metallurgy provided the means to dominate the environment; with the rearing of livestock the hazards and dangers of hunting disappeared. It undoubtedly began a little later than agriculture, around half-way through the 9th millennium before Christ, in Mesopotamia, where the practice of keeping herds of wild **goats** and **sheep**, which were plentiful then, had spread. These animals provided both **meat** and **milk**. According to experts, the goats would have been *Capra aegagrus*, and the sheep moufflon from Iran, *Ovis orientalis*.

The capturing of herds implied **settlement**. Although this has not been proved archaeologically, it remains likely, given that vestiges of cities dating from the 9th millennium have been found on the banks of the Danube and in Anatolia, at Lepenski-Vir in Yugoslavia, and at Catal Höyük in Turkey. However primitive it may seem now, the idea of having herds in an enclosure constituted no less of an invention, and its future testifies to its importance. The segregation of animals led to a **natural selection**, with the animals which reproduced among themselves in the enclosures being shielded from mixing with the much larger genetic stock of the wild herds. Not long after the domestication of the sheep, that of the **dog** took place, but the areas where this began are uncertain. The oldest evidence of the domestication of the dog dates back to 8400 BC, and it came about not in Mesopotamia, nor in the ancient world, but in America, in what is now Idaho state. Nevertheless, it did not take long in the other regions, where the presence of animals and particularly of food scraps could not fail to attract wild dogs.

It is not known exactly when animals began to be artificially selected for their **behavioural** features. The fact remains that the Egyptians used greyhounds for racing and hounds for guarding livestock and houses, and that the Romans did the same with a type of hound.

There are some obscure points concerning the domestication of the dog. Was its ancestor the wolf or the jackal? And should not the appearance of the dog be before the 9th millennium, at least in certain places? Indeed, some rupestrian paintings found in Switzerland dating back 50 000 years portray humans eating animals which greatly resemble dogs. It is possible that the canine species was differentiated from the wolf or the jackal at a very early date and that the first domesticated dogs had in fact been wild dogs. Modern hypotheses nevertheless favour the wolf as ancestor, because it has an identical dental formation to the dog, whilst that of the jackal is quite different.

The following animals were successively domesticated: the **pig**, around 7000 BC; the **cow**, around 6500, in Thessaly and in Anatolia; the **American rat**, around 6000 in Peru; the **silkworm** and the **lama**, around 3500, the first in China (see p. 000), the second in Peru; the **donkey**, around 3000 in Egypt; the **two-humped camel** and the **dromedary**, also around 3000, the first in southern Russia, the second in Saudi Arabia; the **horse**, around 3000 in Ukraine; the **bee**, around 3000 in Egypt; the **banteng**, an Asiatic bovine, around 3000 in Thailand, closely followed by the **Asian buffalo**, around 2500 in the Indus valley; the **duck**, around 2500 in the Near East; the **yak**, another Asiatic bovine, around 2500 in Tibet; **poultry**, around 2000 in the Indus valley; the **cat**, very late, around 1500 in Egypt; the **goose**, its contemporary, in Germany; the **alpaca**, also around 1500 in Peru. The last animal to be domesticated was the **reindeer**, in the 1st millennium, in the Pazyrik valley in Siberia.

The geographical segregation of the species meant that genetic selection was already in operation; then, on the grounds of certain features which tended to appear in certain stocks, the breeders proceeded to carry out **artificial selection** which led to the creation of new species at a very early date. Thus the **hound**, also known as the English dog, appeared in Egypt around the 1st millennium BC; the **chow-chow** appeared in China during the 2nd century BC, and, possibly a descendant, the **Pekinese** appeared during the 8th century. Still on the subject of dogs, the **Roman hound**, now known as the **Rottweiler**, appeared in the 2nd century BC while the **greyhound** was known long beforehand, around the 6th millennium in Egypt. Crossing considerably diversified the varieties within all the species.

It was only in the 18th century that the English first introduced the custom of drawing up **genealogical trees** for horses and cattle.

Market gardening

Anon, Anatolia, around 7000 BC; Rome, 4th century BC

The term 'market gardening', which was originally the **cultivation of vegetables on marshland**, is used here in its broad sense, as opposed to '**cereal cultivation**' (see p. 12). This type of cultivation seems to postdate cereal cultivation by about a thousand years, and it must have proceeded from the gathering of **leguminous plants** and **berries** in the same way that cereal cultivation proceeded from the picking of **wild grasses**. It probably appeared at approximately the same time in Anatolia, South-East Asia, China and Central America, and then a little later in the Andes. It would have been helped by the end of the last glaciation, which occurred 15 000 years before our time. A progressive warming of the climate ensued, which was noticeable from the 12th or 10th millenium, and there was a renewed vigour among vegetation sensitive to the cold, including leguminous plants.

Market gardening constitutes an invention because it consisted of experimenting with removing an edible plant, or one with an edible fruit or berry, and transplanting it on different ground, either directly or by sowing its seeds. Although it proceeds from the same intellectual process as the invention of cereal cultivation, it differs in that the specific growth conditions required by the vegetables are much more complex than those of cereals. One can therefore suppose that its completion needed more agricultural observation, and this is why it took much longer to come about.

It is not known exactly which species were used in the first market gardens. In Anatolia, which seems to have been the most advanced agricultural centre of the time, **lentils, nettle-trees** and **capers** featured among the species grown, some 7000 years before our time. In the New World there were **marrows, pumpkins** and **gourds**, which when dried could also be used as fishing net floats, then, around 5000 years, the **potato, tomato, amaranth** and **capsicum**.

In Egypt between 3200 and 2780 BC, market gardens are known to have been greatly expanded, since they included **asparagus, cardoon, celery, cabbage, lettuce, onions, chick-peas, radish** and **water-melon**. A little later the **cucumber** appeared. It seems that Egyptian market gardening served as an example to the rest of the Mediterranean basin owing to its technical nature and its great variety. By the 1st century BC their products and methods had gone beyond the Mediterranean zone, reaching Asia Minor, Transcaucasia, Persia and Turkmenistan.

In return, trade with these regions enriched Mediterranean market gardening with remote varieties of indigenous vegetables: **aubergine, rhubarb, yam, taro** and **pumpkin** from western China; **broad-bean** from India; **garlic, carrot, melon** and **spinach** from the regions which are now Tadzhikstan and Ouzbekistan; **cabbage** and **leek** from Turkmenistan.

Similar types of trade enriched Egyptian and, indirectly, Mediterranean and European cultivation with African varieties. Still in the 1st century BC, Egypt's market gardens included the following: **okra, artichoke, chives, watercress, chicory, beetroot, parsley, parsnip, rhubarb** and **turnip**, as well as various varieties of **pea** from the Upper Nile regions.

In the 4th century, Rome took over the market gardens and most of the species which had spread over the Mediterranean basin. Paradoxically, a great many of these varieties disappeared and market gardening collapsed at the end of the Empire. The cabbage and the cucumber were reintroduced in Europe only in the 9th century. Spinach, which was under common cultivation in China in the 7th century, did not reach Europe until five centuries later. Several of the species mentioned above which had previously been commonplace in the Roman Empire, such as the artichoke and the aubergine, reappeared in Europe only very much later. In the meantime Europe had been conquered by the potato which had been introduced from South America in the mid-16th century.

Until it was colonized, the New World had only indigenous species. The first vegetables were the carrot and the garden pea, which were introduced into North America at the beginning of the 17th century, followed by the onion. Paradoxically, the potato which has been widely grown in Europe since the 1550s did not begin to spread in North America until the beginning of the 18th century. Trade between the Americas helped the exchange of species for market gardening, often in an indirect way. It was in this way that the melon was introduced to Central America by the Spanish at the beginning of the 16th century and then spread to the southern states of North America a century later, as did the cucumber. Garden peas and carrots were imported to South America at the beginning of the 18th century for the benefit of the colonies, whilst the American Indians continued to eat traditional varieties such as pumpkins, tomatoes, Lima beans and sweet potatoes.

Central and southern America did not progress from market gardening to proper cultivation until the end of the 17th century, with the plants which could be used for it growing wild and abundantly before colonization. The Peruvian memorialist Garcilaso de La Vega, to whom we owe testimonies which are as numerous as they are precise, wrote that the chicory and spinach plants near Lima grew 'as tall as a man' and that the horses had trouble forcing their way through them. It was the eating habits of the Spanish settlers which gradually led to the domestication of wild species such as these, which were so abundant that the same La Vega cites the wild turnips as a scourge.

With the exception of the sweet potato and the taro, Oceania discovered market gardening only at the end of the 18th century when the first Europeans began to colonize this continent. The majority of plants cultivated in Australia and New Zealand, for example, are of European importation. When Europeans arrived in New Zealand in the 18th century there was not even any cereal cultivation there (see p. 12).

There are few areas in agricultural technology that are as irregular as that of market gardening. The best example is given by the tomato (from the Aztec word **tomatl**) which existed in the wild state before the colonization of South America and was imported to Europe in the 16th century as a botanical curiosity (it appears in a description by an Italian author dated 1554 under the name of **pomi d'oro**, or golden apple, because the fruit in its wild state was yellow). Although it was eaten by the South American Indians, who gathered it in the wild, it was looked upon with suspicion by the Europeans until the 18th century. It was the American statesman Jefferson who introduced the cultivation of it into the southern American states in 1781 and it has been in common consumption since 1812. The northern American states, however, refused both the importation and the consumption of the fruit until 1900, for they believed it to be poisonous. The American domesticated varieties nevertheless spread throughout South America at the end of the 19th century. In the meantime the European varieties spread to the Mediterranean and Asia. Thenceforth, hybridization began.

Mill

Anon, Near East, c.2nd century BC

The first use of naturally available energy for the purpose of obtaining food seems to have been the **horizontal wheel water-mill**, which appeared around the 2nd or 1st century BC, probably in the Near East; from there it spread to Greece, then Italy and the rest of Europe. The wheel was submerged in running water which made it turn; it transferred its movement via a fixed axle to an overhanging **flat millstone** situated inside the mill. This type of mill was used for grinding **cereals** or **oil-producing plants**. According to the account left by Vitruvius the **vertical wheel water-mill** was only realized during the 1st century. The horizontal mill could only be driven as a result of the invention of **toothed cogwheels** joined to the **driving wheel** and the **millstone**. According to some authors, **windmills** may have appeared in Syria in the 7th century, but proof of their use dates only from the 10th century. They are found from China to Portugal; in Asia they were also used for **irrigation** and **drainage**.

Probably the original basis for the principle of the propeller (see p. 200), the principle of the mill was quickly to give rise to the sciences of **hydraulics** and **aerodynamics**. Although they were considered obsolete after the Industrial Revolution, the windmill and the water-mill were to be revived in different forms in the second half of the 20th century, for the exploitation of the natural energy of the wind and the tides.

The adoption of the **hydraulic mill** contributed enormously to Europe's industrial and commercial expansion. In 1086, 5624 were recorded in the Domesday Book of England. Water-mills and windmills very soon were used for purposes other than food processing, such as the sawing of wood and the fulling of felt, etc. The mill may also have contributed to the development of **clockmaking**, by providing an improved knowledge of cogwheel systems.

Noria

Anon, Near East, 3rd century BC

Also known as **sakia**, the noria is one invention which has played a large part in the agriculture of the ancient East and Far East, and then in Europe from the Middle Ages, for it enables very effective irrigation in return for a basic installation and a very small amount of **animal energy**. It was invented around the 3rd century BC in the Near East, perhaps in Mesopotamia, and consists essentially of a large horizontal wheel, the vertical teeth of which activate a vertical wheel which is equipped with buckets and half submerged in running water. The buckets empty into an irrigation channel. This type of apparatus can irrigate 3000 m^2 in 12 h if the water from which it draws is 2 m deep, and 1900 m^2 if the water is between 6 and 8 m. The noria is still the fundamental irrigation method of vast regions of Africa and Asia.

Ploughshare

Anon, Egypt, 3000 BC; anon, India, c.600 BC; anon, China, 481 BC; anon, Rome, 200 BC; anon, the Netherlands, around 1600; Ransome, 1785; Deere, 1837

It is impossible to assign an exact date to the origin of the ploughshare, a piece of hard material which cuts the soil horizontally before the planting of **seeds**. It is nearly certain that when the nomadic **Stone Age** peoples began to settle, and the hunting and gathering lifestyle changed to one of cultivation and rearing, roughly between 9000 and 7000 years before our time (Natoufian period, Palestine), man must have understood the need to bury the seeds in order to protect them from animals and the wind. Ninety centuries BC the Natoufians possessed **sickles** — shaped stones attached to shafts — which they probably used to dig and make furrows in the ground.

Such instruments have not been found, however, and the oldest definite description of the ploughshare dates back to Egypt 3000 years BC. It was a piece of fire-hardened wood, resembling a sabre in shape, which was fixed to a wooden frame pulled by a pair of draft animals (probably cows). The relatively soft nature of the soil in the Nile valley enabled this type of work to be carried out without too much effort.

It was still the wooden ploughshare that was found in the Indian subcontinent, dating to between 600 and 300 BC, although it could have existed previously. It consisted of a wooden angle, probably also hardened using fire, which may have been embellished with an iron cutting edge. Linked by a fork to the necks of the oxen which pulled it and extended by a lever which allowed it to be steered, it only cut a furrow in the soil and did not turn it. Some of these ploughshares are light enough to have been carried on the shoulder of the labourer, while others are much heavier. The work also seems to have been carried out using an instrument typical of India, the **kodali**, which looks like an **adze**, and which consists of an iron blade fixed at right angles to a shaft of wood.

Though very advanced in numerous technological sectors, China seems to be behind in the invention of ploughing

It was in Germany, around 1860, that fixed motors first began to be used to drive ploughs from a distance, evidently within a limited range.

Egyptian plough *Thebes, Deir el Medineh, Sennedjem's tomb, 19th dynasty*

implements, for the oldest ploughshare to have been found there dates back only to 481 BC. It was a V-shaped metal point, which presumably was fixed to a wooden frame and pushed by the strength of the arms. There are no records of animals being used to pull this apparatus before the 1st century BC.

The **wheel** is missing from these three examples. In the Americas, until the arrival of Columbus, agriculture was even more primitive since the work was done only with a sharpened stick fixed to a curved shaft, which served as a ploughshare. Once again the strength of the arms was used, and apparently in a very uncomfortable position: kneeling.

It was only around the 16th century that the ploughshare had its first fundamental modification, heralded by the **auricles** with which it had been fitted in 200 BC by the Romans for the purpose of turning the soil. Becoming dissymmetrically shaped, the **Flemish ploughshare** was equipped with a **mould-board** with which it formed one single surface. Its great advantages were that it reduced the effort of traction, was adaptable to hard ground, and turned over the earth, thus dispensing with the need for a helper previously required to break up the lumps of earth and prepare the furrow so that the seeds would fall in correctly. The wheeled ploughshare appeared in Cisalpine Gaul in the 1st

Plough with three ploughshares*, 14th century, Jacques Besson: 'scene of mathematical and mechanical instruments'. Lyon, 1578*

century but evidently followed a similar timescale to the Flemish ploughshare. Nearly three centuries had to pass before the agronomists eventually adopted the Flemish plough, undoubtedly because the ploughshare had the significant inconvenience, from an economical point of view, of being made only in the large forges. It was only in 1783 that the first industrial ploughmaking factory was founded in England.

Further, it was only in 1785 that the Englishman Robert Ransome patented the **cast-iron ploughshare**. In 1837 the American John Deere had the idea of making ploughshares more easily by cutting out a saw blade and fitting it, by hammering, to a piece of shaped wood.

The **coulter**, which first appeared in the 14th century in Europe and was made by an unknown inventor, is the vertical iron soil cutter which is fixed to the beam of the plough. It was kept in use until a late stage, depending on the region.

Systematic crop rotation

Anon, Europe, 3000 BC; Rome, 3rd–2nd century BC

The practice of **crop rotation**, which is intended to stop the soil becoming impoverished, is one of the greatest inventions in agriculture. It was imposed progressively by **settlement**, then by the establishment of **large urban centres**, and seems to have been practised since the 3rd century BC, based on a **winter cultivation**, a **spring cultivation**, and a third **fallow** year. This is known as **three-year rotation of crops** which, by alternation and the use of fertilizer, assured an average yield of cereals (see p. 12). Oddly enough it took many centuries to become widespread, although it was systematically applied throughout the Roman Empire from the 3rd or 2nd century (for a long time the Romans endeavoured to impose **two-year rotation** and seem to have practised it alongside the three-year rotation). It was only in the 12th century that England eventually adopted it, and in the 16th that Russia recognized its importance. This crucial practice thus took nearly five centuries to spread throughout Northern Europe and nine in the East.

It had immense consequences, the most important of which was the extension of **fallow pastures**, which were beneficial to breeding and assured the increased use of the **horse**. A second consequence was the extension of **wheat**, which replaced **millet**, popular until then because it fatigued the soil the least, nearly everywhere. The food supply was therefore altered, and now rested on, as well as wheat, the three most widespread spring crops, **oats**, spring **wheat** and **leguminous plants**. A third, incidental consequence was a better yield from livestock. This can be attributed not only to the extension of pastures and a new abundance of wheat, but also to a false belief of the farmers at that time. When the livestock did not thrive on the pastures, they blamed the weeds and the pastures were limed in order to burn them. Since it was precisely a lack of **calcium** from which the livestock suffered, the lime rectified this deficiency, producing healthier livestock.

Chemistry and Physics

The number of actual inventions made in chemistry and physics up to the end of the 19th century was small. This poverty is nevertheless compensated for by the large number of discoveries, the majority of which have only a brief history, since they began around the 17th century. It was only in 1772 that the Swede Scheele discovered oxygen, for example, and it was not until the 19th century that we began to have some idea of what carbon is! Again, the inventory and understanding of discoveries are hampered by the erring ways of theories. Hardly had they emerged from the mystical fog of alchemy than chemistry and physics got tangled up in philosophy, and the phalanxes of the best specialists vainly endeavoured to harmonize their discoveries with the long-winded postulates of 'phlogistics'. Dalton's atomic theory had the greatest trouble being unanimously accepted, and even at the beginning of the 20th century a chemist of Berthelot's reputation claimed that the formula of water was H_2O_2!

In order to invent something, a correct theory must be used; apart from the intuitions of a genius such as Lavoisier or Faraday, most theories failed, both in chemistry and in physics. They would continue to fail for a long time into the 20th century; Ernst Mach, the famous Austrian physicist, would be seen to affirm, with some irritation, that he could not support the theory of atoms ... and that was in 1922!

In the area of applicable realizations, therefore, chemistry and physics amount merely to formulae or even to chance discoveries. The first discipline to suffer was medicine, which found itself at the mercy of the pharmaceutical industry, and until about 1930 remained in subjugation to pharmaceutical preparations, themselves derived from an improved apothecary. Industry was concerned mainly with reshaping materials, as in metallurgy, being evidently ignorant of all or nearly all crystalline structures and knowing pathetically little about synthetic materials. In the domain of energy, only really electricity was known, and it was not until after the middle of the 20th century that physics would be sufficiently developed to generate electronics.

Chemistry and physics can therefore be considered as the sciences which have been last to develop, and have done so as a result of improvements in theory and new means of microscopic investigation.

Alcohol (synthesis of)

Hennel and Sérullas, 1828; Berthelot, 1854

The synthesis of alcohol, which founded **organic chemistry**, is one of the most singular moments in the history of science and inventions. In 1854 Marcellin Berthelot officially claimed it to be attributed to him, and it remains so, although it had already been done in 1828 by a little-known English chemist, Henry Hennel, and, independently, by the Frenchman G.S. Sérullas. Still worse, this synthesis rested on a set of absurd ideas, which were dominated by Berthelot's rejection of **atomism**, a paradoxical rejection when one knows that chemistry and atomism are inseparable!

Berthelot carried out his synthesis by putting **ethylene** in a flask at ordinary temperature and pressure, together with a quarter of its volume of pure **sulphuric acid** and a few kilos of **mercury** as a catalyst. The flask was shaken violently for a long time. During this operation the ethylene reacted with the sulphuric acid to give what is now called **ethyl hydrogen sulphate** but was at the time called **sulphovinic acid**. At the next stage the sulphovinic acid was extracted and six times its volume of water was added. This mixture was distilled and, after purification, **alcohol** was obtained.

This synthesis was carried out in 1854 and was published in January 1855 in 'Reports from the Academy of Sciences'. It appeared to verify one of Berthelot's theories: that it is possible to reconstitute any substance of organic origin, such as alcohol, which until then had only been obtained by the distillation of wine, from non-organic elements, which is quite correct. This experiment therefore invalidated the theories which still counted on a mysterious '**vital force**' which alone was capable of 'organizing' organic substances. Berthelot, who was then only a chemistry assistant at the College of France (and who tried in vain to profit from the commotion caused by his paper to get himself elected to the Academy of Sciences) had nevertheless achieved a masterstroke by intuition; he did not even know the chemical formula of ethylene, which he believed was C_4H_4 while it is actually C_2H_4, or the formula for water, which according to him was H_2O_2.

The whole business would have proceeded no further if the great chemist Eugène Chevreul, who was then 70 years old and to whom we owe the first **analyses of glyceride**, had not been surprised that Berthelot had not cited his predecessors, as custom dictates. Berthelot replied: 'Whatever speculative opinions there may have been on this subject, no alcohol has ever been produced from **hydrogen carbide**'. Chevreul rebuked him and officially reminded him that both Hennel and Sérullas had independently carried out the same synthesis. Chevreul even remembered that Sérullas's paper on the subject had been presented to the Academy of Sciences on 22 October, 1828, 26 years before Berthelot's experiment. Nonetheless, Hennel and Sérullas were only credited with applying knowledge and ideas which were already 'in the air' at the time. From 1820 it was thought possible to break down alcohol into ethylene and water, and therefore that it was possible to reconstitute it using water and ethylene (the mercury used by Berthelot had little part to play). In 1828, the same year that Hennel and Sérullas synthesized alcohol for the first time, another great chemist and his pupil, Jean-Baptiste Dumas and Pierre F.G. Boullay, had published a theory according to which the derivatives of alcohol consisted of ethylene and other molecules. The **synthesis of acetylene**, which was also claimed by Berthelot a few years later, was to display the same fundamental faults as that of alcohol.

> In his excellent and somewhat scathing biography, *Berthelot, autopsy of a myth*, Jean Jacques writes that Berthelot naively recommended that the flask containing the elements for the reconstitution of alcohol should be shaken '53 000' times. The reaction in fact takes place without being shaken — theatrically — thousands of times.

Aspirin

Anon, 1st century; Stone, 1763; Gerhardt, 1853

Aspirin, the commercial name for **acetylsalicylic acid**, is one of numerous inventions which have been the object of claims which contradict their paternity. It is said that the active ingredient has been known since at least the 1st century. Numerous archaic pharmacopoeial recipes for fevers and headaches feature the decoction of the **bark** of the **willow tree**, **Salix alba**, and the **silver birch tree**, **Betula lenta**, and of the leaves of numerous plant species such as **Spiraea**, which are plants from the Rosaceae family to which **meadowsweet** and **wintergreen**, **Gaultheria procumbens**, belong. These plant species all in fact contain **salicylic acid**, which has a formula and properties almost identical to those of acetylsalicylic acid. This element was only actually 'reinvented' in 1763, when the English doctor Edward Stone gave a paper to the Royal Society of London on the effects of the use of willow tree bark in the treatment of fever. Thus the practice of decocting this product was taken up again and extended. In 1829 the French pharmacist H. Leroux identified the active element in willow bark to be **salicin**.

The modern age of the aspirin began in 1838 when the Italian R. Piria isolated pure salicylic acid for the first time, extracting it from methyl salicylate taken from birch tree bark. The modern inventor of aspirin can also be said to be the Frenchman Charles Frédéric Gerhardt who in 1853 was the first to synthesize acetylsalicylic acid by treating sodium salicylate with acetyl chloride, itself extracted from birch tree bark.

Atomic theory

Dalton, 1803

The notion of the **atom** (from the Greek *a-tomein* which means 'undividable') dates back to the Greek philosopher Leucippus in the 5th century BC, who passed it on to his disciple, Democritus, and seems to have been a success, for Epicurus and his disciples carried it on into the 4th and 3rd centuries. Lucretius, a Roman poet and philosopher, took it up again during the 1st century BC. It can therefore be maintained that it has no single father, having diffused throughout various cultures, and that it is not a proper invention as such, but this would be to neglect the fact that, despite its success, the notion of the atom was metaphysical, since it did not rest on any scientific observation. Even when the Englishman Robert Hooke took it and organized it in the 17th century, and when he conjectured, with astonishing far-sightedness, that the properties of matter must be envisaged in terms of movements and collisions of atoms, it was still no more than a theory. Similarly, when the strange Jesuit, diplomat and scientist Ruggero Boscovitch postulated in the mid-18th century that atoms must be considered not as little round balls but as centres of force, thus prefiguring 20th-century discoveries in a disturbing way, it was still only an intuition, albeit a brilliant one.

No one had any idea, for example, how many atoms there were in a milligram of given material, which was one of the essential elements for the verification of the theory. Then suddenly everything began to happen quickly. In 1792 the outstanding

Dalton was the first to describe the congenital incapacity to distinguish certain colours, known since then under the name of Daltonism (colour-blindness), for the very good reason that both he and his brother were affected by it.

English physicist and chemist, John Dalton, who was a professor at the age of twelve, published a treatise on meteorology, this being his favourite field of scientific investigation. Strangely enough, it was after the publication that he was led to ponder another theory, which had been expressed by Newton to explain certain problems which had been revealed in the study of gases: the theory stated that since a fluid consists of particles, its density is proportional to its pressure, because the force of repulsion which is exerted between the particles is proportional to the distance between them and the centres of the particles. When applied to the air, this theory was difficult to develop, since air is a mixture of gases, a fact unknown to Newton. Dalton overcame this difficulty by supposing that the repulsion was only exerted between atoms of the same gas.

In 1803 a colleague of Dalton's, William Henry, discovered that there was a relation between the pressure to which a gas is subjected and its solubility in water. The discovery struck Dalton, who applied himself to the following task: to establish the relationship between the distances separating the particles of a gas in the natural state and those of a gas in solution. Dalton had the intuition that this relation would vary according to the gas, but he could not know that there was also a relationship between the density and the atomic composition; nevertheless, on 6 September, 1803, he began to formulate a first table of proportional numbers. Two years later, having invented a notation system and determined the weight-volume relations of various gases, Dalton had established the proportional numbers of particles by gases. Thus he founded the scientific bases for the **atomic theory**. His work was continued for the time being by Gay-Lussac, then by Alexander von Humboldt, by the Italian Amedeo Avogadro and by André Marie Ampère. The two most important discoveries made by these great scientists were, in 1808, that of the law according to which gases combine with each other following simple volumetric relations, and that of the fixed relation between the density of a gas and the number of molecules which it contains in a determined volume, which was revealed by Avogadro.

From this time onwards chemical analyses of compounds would rest on precise numerical bases.

Calorimetry

Wilcke, Black, 1760; Lavoisier and Laplace, 1780; Thompson, 1798; Hermann, 1834; Herschel, 1847; Favre and Silbermann, 1850; Bunsen, 1870

The concept of heat, which is essential to chemistry and physics, remained shrouded in mystery until the last quarter of the 18th century. It was the first attempts to measure heat that led to an understanding of it. From 1697 a great many scientists supported **phlogistic theory**, which was expounded that year by the German chemist and physicist Georg Ernst Stahl. This theory postulated the existence of an element, the phlogiston, which had neither weight nor mass and which was found in elements which burned easily. From this, the German Carl Johan Wilcke and the Englishman Joseph Black devised, independently and in the same year (1760), the **caloric theory**. According to them an indestructible fluid called the caloric accumulated when a body was heated and subsided when it was cooled down. This obviously false theory, which was approved by numerous scientists until 1824, had a certain scientific advantage nevertheless, because it led Black to lay the foundations of calorimetry. He had the idea of defining the caloric units which would be necessary to raise the temperature of a body. Here Black made the fundamental and correct distinction between heat and temperature, and found that the same quantity of heat raises the temperature of wood and of metal by different amounts. Thermometers, which had existed since at least the 17th century (see p. 172), were not capable of registering the quantity of heat present in a body, and therefore Wilcke and Black can be said to have put the science on the path of a fundamental modern discipline.

In 1780 Antoine Laurent de Lavoisier and Pierre Simon de Laplace, refuting the phlogistic doctrine, invented an original device to measure the quantities of heat generated or released by bodies in certain conditions. Their apparatus consisted of a test-tube sealed in ice and resting on a bath of mercury inside a ball, the bottom end of which opened on to a graduated scale. The quantity of heat released in the tube was proportional to the mass of ice which melted and then pushed down the mercury. This was the first known calorimeter, the **ice calorimeter**. The principle was inspired by the theory of Wilcke and Black, but Lavoisier and Laplace were not aiming to found calorimetry; their interest was in the study of the phenomena of respiration and combustion. The instrument was successively improved through the reduction of the calorific loss by the German Jakob Hermann in 1834, the Englishman John Herschel in 1847 and the German Robert Bunsen in 1870. In 1850 the Frenchmen Pierre Favre and Jean-Théophile Silbermann modified the principle of Lavoisier and Laplace by removing the ice and inventing a **mercury expansion calorimeter**.

Interestingly, it was a previous observation which led to the discovery of another very important principle in calorimetry. In 1798 the Anglo-American Benjamin Thompson, Count Rumford, was struck by the great heat given out by a cannon because of friction. He then devised a remarkable experiment: he set up a metal cylinder to be rubbed by a mechanism activated by two horses; the cylinder was submerged in a caisson filled with 8 l of water. After two and a half hours the water reached boiling point. Thompson was thus the first to have demonstrated an equivalence between mechanical work and the heat produced.

By 1843 the Englishman James Prescott Joule had already established the numerical relationship between a given quantity of electricity and the amount of heat that it produced. In 1849, after six years of research, he determined the precise numerical relation which was first established by Benjamin Thompson.

The calorimeter brought about considerable progress in physics and enabled Sadi Carnot to develop the theory of **thermodynamics** in his *Réflexions sur la puissance motrice du feu* (Reflections on the driving power of fire) (1824). It allowed the recognition of the calorific conductivity of materials, leading to the **physics of gases** and **molecular physics**. It also enabled considerable progress to be made in **physiology** which would eventually lay down the foundations of a science of the body's energy use. From the moment when the norms of human needs were seen in terms of calories according to age, weight and type of activity, the foundations of **nutrition** could also be laid. It was not long before **dietetics** followed.

Distillation

Anon, 4000 or 5000 BC

The **separation of a liquid by boiling** is an extremely old process, although the date and place of its invention are not known. It was practised in very ancient times for the extraction of **alcohol** from the wort of fermented drinks, by placing a piece of material over the pot of boiling liquid which would then become impregnated with alcoholic vapours; the alcohol was recovered by torsion. This process seems to have been in use in the Near East and in Asia four to five millennia before Christ. The **still** does not seem to have appeared until after the year 1000, and this is generally supposed to have been in Italy around the 12th century. The first stills were made of metal.

Explosives

Anon, China, 2nd century BC and 9th century; Sobrero, 1846

Until about the last third of the 19th century no explosive had been made for military purposes, quite a remarkable fact. The first known explosive appeared in China around the 2nd century BC; it consisted simply of sections of **bamboo** which were thrown into a fire and which exploded because the air inside was suddenly heated. It was apparently only used for festivities because of the noise it made and because it gathered the people of the villages together. It was also for festivities, in this case for **fireworks**, that in the 9th century the Chinese also invented **gunpowder** (see p. 220). This did not reach Europe until much later.

Until around 1788 gunpowder was the only explosive which was known and used. During that year interest arose in **chlorates** and **perchlorates**, salts of the chloric and perchloric acids, particularly **potassium chlorate**. Attempts were made to manufacture it industrially, in Europe and the United States, but the accidents which ensued put an end to these.

The first modern explosive, **nitroglycerine**, was derived from research, albeit fruitless, on **textiles**. It was invented by the Italian Ascanio Sobrero. Dynamite, which was soaked up by sticks of porous silica or kieselguhr, hence its name **guhr dynamite**, or sticks of charcoal, was first made for use in civil engineering by the Swede Alfred Nobel; it actually only consisted of an adaptation of nitroglycerine. In 1867 Nobel made **straight dynamites**; these were less dangerous to handle and less susceptible to the cold which reduced their detonating power. Instead of having a pure nitroglycerine base, they included **polyhydric alcohols** and *sugars* mixed with **wood pulp**.

***One way of setting up a boat as a firework display**. 'The pyrotechny of Hanzelzt Lonain'. Pont-à-Mousson, 1630*

Galvanoplasty

Jacobi, 1837

Galvanoplasty is the set of techniques which use electrolysis to deposit a layer of metal on another metal or other material. Its origin is relatively obscure.

It is known that it was the German Moritz Hermann von Jacobi who officially invented galvanoplasty in 1837. The method was one of disconcerting simplicity, for it consisted of submerging the object to be covered in a bath, in which the electrolyte was the salt of the metal to be deposited. The object was a conductor of electricity and thus played the role of cathode. His first undertaking was a **silver-plating**, which was carried out in Russia and for which he received the estimable Demidof prize. The process rapidly became international, for it allowed medals and coins made of copper to be coated in silver, for example, copies to be made, and the creation of the silver-plate industry. In the long run, it led to the **nickel-plating** and then the **chrome-plating** of mechanical parts.

In reality, galvanoplasty had been discovered 37 years previously, practically at the time of the manufacture of the first **electric battery** by the Italian Alessandro Volta, when it was suddenly realized that the strips of zinc deteriorated rapidly when they were mixed with strips of copper. In 1801 it was established that if a piece of zinc, copper or silver was deposited on the cathode, it would become soldered to it. It can therefore be suggested that the premises of galvanoplasty were set down by numerous physicists who, after the voltaic battery had been discovered, made use of it to carry out various experiments.

It can also be said, however, that the origins of galvanoplasty are even older, and that they go back to the 16th century. At this time the alchemists noticed that an iron rod decomposed when it was soaked in '**blue vitriol**', that is, a copper sulphate solution; the iron was then rediscovered as a ferruginous deposit (this was the beginning of the end of illusions about transmutation). This experiment demonstrated that in fact it was possible to dissolve a metal in a liquid. It is possible to take the process of galvanoplasty back even further, for example to the electric battery of Parthian origin which was discovered in 1957 among the artefacts kept in the museum of Baghdad, and which has been dated to 224–250 BC; it might have been used to supply electricity for silver-plating processes, for example, that is if one is inclined to speculate.

> If one wishes to go deeper into the hypothesis of a very ancient practice of galvanoplasty, there is no satisfactory explanation of gold- and silver-plating on some ancient objects. This plating is sometimes so thin that it seems unlikely it could have been achieved by systematic hammering.

Nitric acid (preparation of)

Geber (?), 8th century; Glauber (?), 1648; Kuhlmann, 1838

The origins of the preparation of nitric acid are obscure. It is not actually known who was the first to find it. Some authors attribute the paternity of the invention to the Arab Geber, in the 8th century, but there is no absolute proof. Geber could in fact have benefited from previous knowledge. The fact remains that he is the source of knowledge on the subject which was used by **alchemists** throughout the Middle Ages. This **strong monoacid**, HNO_3, was used in research on the **transmutation of**

metals. It attacks all metals except for gold and platinum, and produces **nitrates** at the end of the reaction. Known under the names of *Aqua fortis*, *Aqua dissolutiva*, *Aqua prima* and then *Spiritus acidus nitri* or *Spiritus fumans Glauberi*, one of its first practical uses was in aqua fortis **etching**. It is thought that the method of manufacture, which remained classical for a long time, was established in 1648 by the German Johann Rudolf Glauber; it consisted of heating **potassium nitrate** with concentrated **sulphuric acid**. In 1838 the Frenchman Charles Frédéric Kuhlmann invented a new method of preparation — the oxidation of **ammonia gas** in the presence of **platinum**. This process is still used, although it is tending to be gradually replaced by the one invented in 1901 by the German Carl Bosch, which consists of using a catalyst of ferric oxide activated by manganese and bismuth.

Nitroglycerine

Sobrero, 1846

After the arrival of gunpowder, the **technology of explosives** made little progress. It only progressed in the 19th century by means of research in the domain of **textiles**. It was actually in endeavouring to make an **artificial fibre** which could be woven out of a **cellulose paste** that the German Christian Friedrich Schönbein ended up with **nitrocellulose**; this new product, obtained by treating cellulose with nitric acid, did not have satisfactory textile properties because it was highly inflammable. That same year, 1846, the Italian Ascanio Sobrero, who was interested in the inflammable properties of nitrocellulose, treated **glycerol**, a by-product of soap and candle manufacture, with **sulphuric and nitric acids** at room temperature. He obtained trinitroglycerine, which is more commonly known as **nitroglycerine**. The preparation of nitroglycerine involved a dangerous, highly exothermic reaction, and the product itself was hazardous since it would explode when knocked. In 1862 the Swede Alfred Nobel invented a means of producing it industrially, by soaking up the nitroglycerine using sticks of porous silica; in this way he made **dynamite**.

Nitroglycerine, a product of organic chemistry, later found a place in pharmacopoeias for the **prevention of angina**.

Soda (extraction of)

Anon, Egypt, c.16th century BC; Leblanc, 1789

The Industrial Revolution increased the demand for numerous chemical substances, including soda. Soda was in fact already necessary for the manufacture of products such as hard **soaps** and certain **papers**, in processes such as **mercerization**, as well as in all **dehydration** operations. Soda is a compound of the metal sodium. The most common form of sodium is its everyday composite, **sodium chloride**, or salt. The extraction of soda, however, is more complex.

In the 16th century BC the Egyptians, who had very advanced glass-making techniques, used native **sodium carbonate**, an **alkali** (from an Arab name, since soda and potash were first made in Arab countries). The Egyptians, followed by other peoples of the Near and Middle East, also seem to have extracted soda from the ashes of plants which they washed in sieves; once the water had evaporated, by exposure to the sun or heating, soda remained. In 1789 the Frenchman Nicolas Leblanc had invented a process for extracting soda, which consisted of combining sodium chloride with **lead monoxide** or **chalk** to obtain a damp substance which was allowed to mature. After a few days the soda appeared on the surface by precipitation, and all that remained was to collect it.

Another process consisted of heating the sodium chloride together with **sulphuric acid**, to produce **sodium sulphate** (provided that the hydrochloric acid vapours had cleared). The sodium sulphate was then heated with chalk, coal or lead monoxide and a black ash was obtained which consisted mainly of **sodium carbonate** and **calcium sulphate**. Finally, the sodium carbonate was dissolved in water and crystallized in order to have it in its commercial form. This was a simple method which became very widespread, but it was costly because it required sulphuric acid which was expensive then. Besides, the Leblanc process had two big disadvantages. The first was the discharge of **hydrochloric acid vapours** into the air, which would pose serious problems as soon as it came to be exploited industrially, since it would require the use of condensation towers. The second disadvantage was the problem of the separation of the soda from the calcium sulphate; this was done initially by reducing the sulphate by heating it with coal and iron filings, which was expensive and difficult for the workers. Then, at the beginning of the 19th century, it was treated with **carbonic acid**, which resulted in the production of **hydrogen sulphate** and, finally, of **sulphuric acid**, a profitable operation, given the price of this acid.

Communication, Culture and the Media

From ancient times until the end of the 19th century communication between humans, the cement of society, saw only two major developments: the advent of printing using movable characters, which was borrowed from China and shaped by the industrious Gutenberg, and electricity. The former enabled the diffusion of knowledge, the latter allowed it to spread at great speed.

After the appearance of modern printing, knowledge which was previously rather esoteric began to become more democratic. It was only in the 19th century, however, that the press was to provide a true reflection of society. It took three centuries to create a literate public for the press. In its democratization knowledge was fruitful too. For a few pence anyone could buy a handbook on chemistry or electricity.

At first this seemed to benefit only the press, but in fact the almost instantaneous exchange of information from one end of a country to the other, then between countries and soon between continents, altered our whole perception of the world. At first the phenomenon was imperceptible, then it became widespread as one of the dominant features of modern times. The Russian or the Englishman, and soon the African or the American, all lost their status as foreigners to each other. An embryonic and somewhat delicate awareness of the unity of the human race emerged from the darkness.

The country became urbanized, crime spread and so lost its exceptional character. The police became a part of daily life. This gave rise to the detective novel and to Poe's invention of the first fictional detective, Monsieur Dupin, who is direct ancestor to Sherlock Holmes and Hercule Poirot. Newspaper owners wanted to increase their readership, and what better way to make the reader buy the following day's edition than to offer him the sequel of a story, preferably a story of action and adventure? The serialized novel was born.

In the same way that the photograph captures the event, so the phonograph captures the human voice. People became accustomed to seeing and hearing what was dead and gone, and, paradoxically, the notion of eternity faded. The flow of communication ended up by creating the feeling that knowledge is transitory.

Alphabet for the blind

Haüy, 1793; Braille, 1825; Lucas, 1837

The first to have the idea of **writing in relief** for blind people was the Parisian calligrapher Valentin Haüy. In 1793 he designed a simplified italic character in embossed print for the use of the blind people at the institution which he had founded in 1785. It had only limited success. The Englishman T.M. Lucas, inspired by stenography (see p. 50), took up the idea and created an **alphabet of phonetic symbols** in relief, using which, in 1837, he published a transcription of the New Testament. His compatriot James H. Frere improved the system considerably by inventing the return line, which meant that the text was followed from right to left and then from left to right so that the finger did not lose its place in the text. Louis Braille, a pupil at the Haüy Institute for young blind people, was 16 years old when in 1825 he devised an entirely new system which allowed blind people not only to read but to write as well. He made his system from signs using twelve dots, which was elaborated by the artillery-man Charles Barber during the wars of the Empire and then simplified and replaced by signs of six dots. The Braille system was not adopted until after 1850 and came into widespread use the day after the 1878 International Congress in Paris.

The advantage of the Braille system, in which each combination of dots (from 1 to 6) constitutes a sign, is the convenience of writing; it is done using a simple awl and does not require a machine. Correspondence between blind people was therefore made possible. During the 20th century the bulkiness of the books printed in Braille gave rise to the creation of electronic machines which can store long texts in a small space and then transcribe them for the reader on to a tactile surface which unrolls as it is needed.

Cinema

Plateau, Stämpfer, 1832; Janssen, 1874; Edison, 1891; Lumière, 1895

As with numerous other inventions, the invention of the cinema took place in stages. It started with the invention by the English physiologist Paris of a cardboard disc on which there were two different pictures. For example, on one side there would be a fire, and on the other a flame; persistence of vision ensured that when the disc was spun quickly on a diametral axle it appeared as if the flame was dancing in the fire. The idea was taken up or 'rediscovered' in 1832 by the Belgian Joseph Plateau, who made a disc with holes around the edge and, on one side, images representing a movement, such as a girl skipping. The disc was placed in front of a mirror and when it was turned on its axle a reconstituted, nearly continuous movement could be seen through the vents as they went round. Plateau called his disc a **phenakistoscope**, whereas the Austrian

Edison's Kinetoscopes soon included a sound accompaniment which has led to the rather grandiloquent claim that the talking cinema has in fact existed since 1895. In 1899 the Frenchman Auguste Baron took up the idea and presented it to the Academy of Sciences. In both cases the sound consisted simply of a synchronized record, usually of music, being played on a gramophone ...

Stämpfer, who invented it in the same year, called it a **stroboscope**. This principle, which was realized in many different ways and was named, depending on the inventor, **pharmatrope** (Henry Renno Heyl), **Zoëtrope** (Horner) and **Praxinoscope** (Emile Reynaud), only really progressed towards the cinema as we know it as a result of two inventions, one the work of the French astronomer Jules Janssen in 1874, and the other American Thomas Edison's **Kinetoscope**, which dates from 1891. The former meant that for the first time a large number of successive pictures of a moving object could be taken.

The principle of the sound-track of the talking cinema was invented in 1900 by the Russian Joseph Poliakoff; it consisted of a track which was read optically by a selenium cell and was well ahead of its time.

The first were of the planet Venus passing in front of the Sun, of which Janssen obtained 48 pictures in 72 s, a remarkable feat. The latter was the first to display a chain of successive images on **celluloid tape**, thanks to Reynaud's completion of it. Edison's funfair Kinetoscopes were to have lasting success until 1909, for they were still in use long after the arrival of proper cinema.

The cinema as we know it today was only born when the brothers Georges and Louis Lumière combined the inventions of their predecessors. In particular they managed to ensure a regular unravelling of the celluloid tape which carried the images by making **perforations** in the film which was held by cogwheels. The 'official' launch took place at a showing at the Grand Café on the Capucines Boulevard in Paris on 22 March, 1895.

Colour printing

Anon, China, c.9th century; Gutenberg, 1437; Schöffer, 1457; Le Blon, 1719; Senefelder, around 1818; Baxter, 1834; anon, c.1850

The origin of colour printing is obscure and, judging from the **playing cards** printed in several colours and dating from the 9th century, the technique must have begun in China. The exact process used is not known but it is thought to have consisted of applying different coloured inks to a block of **engraved wood**, which then simply had to go to the press. Without doubt this was the process adopted by Gutenberg to achieve the letters printed in red in the Bible 'of 42 lines', the first book to be printed using movable characters (some other initial letters were coloured by hand). The Mainz printers subsequently produced books with initial letters in two colours, and the large initial letters of the psalter printed in 1457 by Peter Schöffer comprised complex red and blue patterns. The secret of these letters was discovered in 1830 by the Englishman William Congreve: the initial consisted of two parts of wood or metal which fitted into each other rather like a puzzle; these were inked separately and then carefully adjusted. This was the process adopted in 1486 by the famous English printer from St Albans who produced coats of arms printed in four colours, blue, red, brown and yellow, with the yellow being added separately by paintbrush.

At the beginning of the 16th century a new technique of **wood-cutting** appeared. This was derived from the earlier technique and became known as **chiaroscuro**; it is not known whether it began in Germany or Italy or who invented it. It consisted of making a cutting with strong lines and different coloured spaces in between, with the black lines acting as a partition between the colours and therefore absorbing any possible smudges. At the same time the technique of **stencilling** appeared, which was used mainly in the production of playing cards. In 1630 the technique was extended to **line-engraving** (**aqua fortis** was not used for etching until a century later). The results depended on the talent of the engraver. It could be said that there has nevertheless been no fundamental invention in colour printing since that of Schoiffer.

In 1719, however, Johann Christoff Le Blon, a German of French origin, patented a process (undoubtedly invented previously) which was remarkable for its prefiguration of the theories of **optics**, since it was based on the use of **primary colours** for the reconstruction of the spectrum. Le Blon used it for the reproduction of masterpieces in painting.

It was in this way that he made several plates of the same work, attending to the exact location of the four colours and to their exact superimposition; each plate was worked in **mezzotint** according to a primary colour, and the superimposition of blue, yellow and red in the printings allowed the original colours of the work to be reconstituted (besides taking one print in black). The superimposition of blue and red therefore produced violet; that of blue and yellow, green; that of red, yellow and black gave brown, etc.

The intensity of each colour in the final printing depended on the density of the mezzotint frame, with the tightly packed ones producing dark colours, and the faint ones the opposite. This process was published in France in 1756 and was extremely successful. It was based on the use of **transparent inks** which were the only ones which could produce the required effect. It will be noted in passing that this effect closely resembles that of **four-colour printing using zinc plates** which arose in the 19th century, with the squared mezzotint frame being very similar to the latticed screen of photographic plates.

Around 1818 Aloys Senefelder extended the three- or four-colour process to his invention of **lithography** (see p. 43). This was the birth of **chromolithography**. Some printers, such as the Englishman William Savage, achieved remarkable variety by incorporating 30 different colours into engravings by etching, chiaroscuro or chromolithography. Others, such as the Englishman George Baxter, skilfully combined engraving on steel for the black lines and wood-cutting for the colours.

The development of **photography** in 1858 made the mechanical reproduction of colours possible. This was achieved by making negatives on zinc of the four plates of primary colours to be printed, each negative being made from a colour filter of the same image. This was patented by the inventor Louis Ducos du Hauron in 1858. It should be noted that zinc had already been used for engraving — this was first done by the Austrian Joseph Trentsenky in 1822 — and that **zincography** had been commercialized as a substitute for lithography by Senefelder's colleague, Knecht.

Detective novel

Poe, 1841

The detective novel is the only literary genre comprising specific rules which has been created in modern times. It rapidly became internationally successful. All historians agree on the designation of the American writer Edgar Allan Poe as its inventor. The first ever detective story, *The Murders in the Rue Morgue*, which appeared in the Philadelphian *Graham's Magazine*, actually fixed the first rule of the genre, which is to set out a detective puzzle and to resolve it in the course of the story by **investigation** and **deduction**. A whole literary genre was to be inspired by it, from Agatha Christie to Georges Simenon.

The premises of the detective story are found in some of the stories of the *Thousand and One Nights*, as well as in the famous chapter of Voltaire's *Zadig* where a philosopher describes a lost horse and dog which he has never seen. But Poe remains as the founder of the detective novel, for three other tales, *The Gold-Bug*, *The Mystery of Marie Roget* and *The Purloined Letter*, reflect the aim to constitute a specific genre. Although he never used the term '**detective**', Poe nonetheless shaped the first ever detective, Monsieur C. Auguste Dupin. The term detective was first used by one of the most famous of Poe's followers, Charles Dickens.

Encyclopedia

Varro, c. 47 BC; Pliny the Elder, c. AD 17; Panckoucke, 1780

The idea of an encyclopedia goes back to the first projects on the **rational classification of knowledge**; its very etymology shows that it began with the Greeks during the four centuries before Christ: **en kyklos paideia** means instruction — **paideia** — by a complete or circular system — **en kyklos**. Democritus, Plato, Speusippus, the Sophists, Aristotle and others would have written encyclopedias which have not survived. These works apparently comprised a significant section of philosophical considerations, with the authors conceiving knowledge only within a philosophical system of interpreting the world.

The encyclopedia as a collection of knowledge at once multidisciplinary and non-philosophical, in the modern sense, was first devised by the Romans, which reflects their practicality. The first among the Romans to engage in the writing of an encyclopedia was none other than the most illustrious of all scholars of the time, Marcus Terentius Varro, who according to Quintilian was 'the best informed of all Romans', author of no less than 640 works. The fragments of those which have survived show that some were real thematic encyclopedias in the modern sense of the term; dealing with history as well as grammar, logic, rhetoric, arithmetic, geometry, astronomy, music, medicine and architecture (according to the nine muses), and etymology.

Varro's direct successor was Pliny the Elder around the year 17. His *Natural History* comprised 37 books and 2493 chapters, tackling subjects which were neglected by Varro, such as the actual organization of knowledge or epistemology, cosmography, meteorology, geography, anthropology, physiology, zoology, pharmacy and mineralogy. Although Varro is certainly the inventor of the encyclopedic formula, Pliny the Elder is the one who gave it its modern scope.

Isidore of Seville revived the encyclopedic principle, which had been lost by St Augustin and St Jerome in a re-interpretation of knowledge according to Christian philosophy. After his work at the beginning of the 7th century with the exception of possible setbacks in the systems of philosophical or confessional interpretation, encyclopedias increased in number, from the *Didascalion* by Hugues de St Victor written around 1100, which enriched the set of themes in use with chapters on weaving, armament, navigation, hunting and even the theatre, and touched on methods of exegesis and magic, to the 5020 volumes prepared for the Chinese emperor K'anghsi (Kangxi) by an army of scholars in 1722.

The closest ancestor of contemporary encyclopedias is undeniably the *Encyclopedia* by Jean Le Rond d' Alembert, Jean-Paul de Gua de Malves and Denis Diderot, as well as Jean-Jacques Rousseau, François Marie Voltaire, Charles de Montesquieu, Claude Adrien Helvetius, the abbot of Prades and others. It was begun in 1743 as a translation of the *Cyclopaedia* by the Englishman Ephraïm Chambers but rapidly became the major event of the century of Enlightenment which laid the foundations for the American Revolution and then the French Revolution. When it succeeded in being published in 1772 its 28 volumes represented the first collection of knowledge which was both factual and interpretive of modern times; in rejecting metaphysical fiction in favour of absolute objectivity, and assuming the intrinsic interpretive character of any statement of knowledge, its authors inaugurated the critical method which thenceforth impregnated the whole of scientific culture. It is known that the Jesuits and the police resisted this undertaking, seeing it as 'guilty' of fundamentally secularizing knowledge and removing it from the already established ascendancy of philosophy. In fact, the Encyclopedists did not consecrate any article to philosophy, but treated them under headings corresponding to the schools of philosophy, which immediately invalidated the transcendent role of philosophy in the interpretation of knowledge and prefigured the arrival of **modern epistemology**.

Although they obviously did not invent but masterfully reshaped the encyclopedic formula invented by Varro, the Encyclopedists are inseparable from any history of encyclopedias. The fact remains that Charles Joseph Panckoucke, a key figure but a minor spirit, who in 1780 obtained the right to publish the reviewed and enlarged edition of the *Encyclopedia* in volumes, formally invented the **thematic encyclopedia**. In fact, he divided the 44 parts of the work into 51 separate **dictionaries** totalling 140 volumes. He thus began the formula of the modern dictionary and of publication in volumes. Another innovation was the **universal vocabulary** added by Panckoucke, which in fact constituted the first **reference index** of all the terms used in the encyclopedia. The work was not finished until 1832.

Linguistics

Anon, Greece, 5th century BC; Aristotle, 4th century BC; Dionysus Thrax, 1st century BC; Apollonius Dyskolos, 2nd century; Priscian, 6th century; Jones, 1786; Humboldt, 1821; Grimm, 1822; Saussure, 1879

Linguistics should be considered as an invention insofar as it is an intellectual tool which has enabled the **structures of language** to be studied. It therefore rules over the study of **logic** just as much as it does over efforts to produce **programs** which can teach computers to translate texts from one language to another and to form correct, logical sentences on the grammatical plane; this latter aspect of linguistics, which is linked to logic and computing, is expounded in the second volume of the present work.

Linguistics can be said to have been born in the 5th century BC in Greece, resulting from debates of a philosophical nature aiming to establish whether language is a natural or a conventional activity. Basically this debate set down the essential foundation of linguistics which has dominated its development up to the present day. The Greek **naturalists** maintained that words are formed by **onomatopoeia**, that is, they are constructed with the help of evocative sounds. Hence for us in English, 'Mummy', 'hubbub' or 'cuckoo'. The **conventionalists** disagreed on the grounds that onomatopoeic words were not the same from one language to the next and that they could be removed from a language without affecting its communication potential. Debates of this sort were to direct linguistics towards **word forms** and **grammatical structures**.

A century later a new debate came to light, which aimed this time to establish whether the grammatical forms were regular, which was the opinion of the analogists, or irregular, the opinion of the anomalists. Aristotle and the Alexandrian School were among the former, and the Stoics among the latter. This second debate orientated linguistics towards the study of principal categories or **grammatical structures**, which was the work of the Stoics, and to that of principal **inflectional structures**, which was the work of the Alexandrians; together they formed comparative linguistics. From the 1st century BC onwards the debate was dominated by a work which remained as the beacon of linguistics for a thousand years, the *Technê Grammatitê*. This was a Greek grammar by Dionysus Thrax, an Alexandrian, who defined the inflections of Greek words. In the 2nd century Apollonius Dyskolos in a way completed this classic by adding a treatise on Greek syntax.

With the structures of Greek and Latin being similar, the two works had an immense effect, and served not only to teach these languages but also to codify the use of previous 'common' languages. In this respect it can be maintained that the grammarian Priscian in the 6th century laid the preliminary foundations, for he was at the root of the use of linguistic outlines for the study of other languages.

Modern linguistics or **general linguistics** did not arise until the 18th century when the English orientalist William Jones revealed the striking similarities which existed between numerous words in Sanskrit, the language of India, and their Latin and Greek equivalents. Jones consequently formed the hypothesis of an **Indian source of European languages** which has since been verified and which has served to create the concept of Indo-European languages. This work, which Jones had enriched with considerations on old Germanic languages, Gothic, high German or Hochdeutsch and Old Norse, was continued by the Dane Jacob Grimm, who proceeded in the same direction. Thus the idea of interlinguistic structures began to spread, for example the Sanskrit word *pados* corresponded to the Latin *pedis* and the Greek *podos*, all three words having the same meaning, 'foot'.

The **law of derivatives** ensued from the work of Jones and Grimm. According to this law, words transferred from one language to another followed regular patterns. Suddenly it became possible to discover the original term, going even beyond Sanskrit, and so into Proto-Indo–European. In 1821 the Prussian Wilhelm von Humboldt made considerable progress in linguistics with the idea that there are two forms of language, one 'external', consisting of sounds, and the other 'internal', consisting of grammatical structures which confer a specific sense to the sounds and therefore effects the differentiation between languages. In this way a key concept of modern linguistics began to take shape, that of **structuralism**. Hum-

The study of linguistics has greatly influenced philosophy and psychology since the end of the 19th century. Its theoreticians are placed half-way between inventors and discoverers; they have been discoverers insofar as they have brought a law to light, but inventors insofar as they have applied linguistic structuralism to other areas. Primordially an academic discipline, linguistics was subsequently to prove itself to be a precious technical tool. **Exegesis**, **pedagogy** and the teaching of foreign languages in particular, **philosophy**, **ethnology**, **logic** and **computing** have all benefited from this tool.

boldt had a follower whose fame now sometimes tends to overshadow him. This was the Swiss Ferdinand de Saussure who distinguished on the one hand between the **language system** and **speech**, and on the other, between the **form or signifier** and the **substance or signified**. Speech appeared as a kind of behavioural form of language, independent of the structure of language.

Lithography

Senefelder, 1796

The inventor of lithography was Aloys Senefelder, a dramatist and actor born in Prague in 1771 and living in Munich. Being poverty-stricken and unable to pay for his plays to be printed, he resolved to print them himself by engraving them on copper. This was a laborious process, for it involved both writing back-to-front and the polishing of plates which had already been used. It was also costly, because the copper plates were relatively expensive.

Chance came to the aid of Senefelder. There happened to be a piece of **Kelheim stone** on his table, the kind used for the preparation of printing ink, when his mother urgently needed to write out a laundry bill. The laundress was waiting, and Senefelder, having neither paper nor pencil to hand, wrote the account on the Kelheim stone. The ink he used was based on wax, soap and lamp-black. When the account was done, Senefelder was about to erase the figures when he had the idea of trying the *aqua fortis* (nitric acid), on the stone [which he had been using for etching on the copper]. He therefore put a border of wax around the stone and covered it with diluted aqua fortis, at the ratio of one part aqua fortis to ten parts water. After five minutes he noticed that the aqua fortis had actually eaten into the stone, leaving the letters and figures which were written on it standing out in light relief, being protected by the greasy ink. He therefore began to ink the embossed part with a printer's pad, thus inventing a new tool which was made simply from a plate covered in fine cloth. This was in 1796; Senefelder had just made a new printing process.

Subsequently it appeared that dampening of the sunken parts was a fundamental condition of lithography, since it prevented any overflow of ink from the inking pad or roller. Still later, Senefelder resolved the problem of the layout: instead of writing the wrong way round, which was slow and inconvenient, he found that it was possible to do it the right way round on a page which was then transferred directly on to the stone, which with a few alterations, would then be ready for the application of the aqua fortis.

The process was internationally successful. It was introduced in Paris in 1802 by Pierre-Frédéric André. In 1836 the printer Auguste Dupont discovered a quarry of stone very similar to the one on which Senefelder had realized his lithographies and he thus spared French lithographic art from expensive importations of Kelheim stone.

Senefelder quickly became an expert on lithography and contributed to the development of this technique to its present point of sophistication. Having become famous in Munich and received additional commercial advantages, in 1801 he was rewarded with a generous pension from the King of Bavaria.

Magic lantern

Shao Ong, 121 BC; Bate, 1634; Faraday, 1836

For a long time the magic lantern was thought to be a reasonably modern invention, which was variously attributed and remembered under the name of **Zoëtrope**. Its origin is much more ancient, however, for descriptions of showings organized in 121 BC by the Chinese Shao Ong for his emperor exist. It is possible that the invention took place even earlier, as suggested by the description of a type of **kaleidoscope** owned by another emperor in 207 BC. The fact remains that the description of the object invented by Shao Ong is the one which corresponds most satisfactorily to our conception of the magic lantern. In 1634 the Englishman John Bate gave details of the instrument in his *Mysteries of nature and art*, without really attributing the invention of it. It was his compatriot Michael Faraday who reinvented the magic lantern two centuries later in 1836. The original instrument seems to have consisted of cut-out paper figurines which were mounted on a merry-go-round and paraded in front of a candle inside a dark room.

Newspaper

Julius Caesar, 59 BC; anon, England, 1590–1610; Verhoeven, 1605 or 1621

The first regular informative newspapers for public use are believed to have been created in 59 BC by Julius Caesar. They consisted of handwritten parchment sheets which were copied several times and posted on to the main buildings in Rome and in the principal provincial towns; they were called *acta diurna* and reported on edicts, events, outcomes of battles, nominations, judgments, executions, naval and military news, births, marriages and deaths. A more restricted form of this existed in China between 618 and 907; it consisted of a court newspaper or **pao** (bao), duly printed in several tens of copies and publishing the official news; this kind of newspaper lasted until 1911. Between 1590 and 1610 a more or less regular newspaper with much the same content appeared in London, the *Mercurius Gallobelgicus*, which mainly gave news of the continent. The *Nieuwe Tijdingen*, which is thought to have been launched around 1605 in Anvers by the Dutchman Abraham Verhoeven (but the oldest available copy dates to only 1621) was undoubtedly the first real ancestor of the modern press, both because of its greater frequency and regularity of publication and the variety of its information. It seems to have been an improved form of a commercial newspaper, *De Courante Bladen*, which existed previously and which was published for the tradesmen of Anvers and Venice. It must be added that the high price of the first newspapers meant that for a long time one went to read them in a café or a club, and that consequently they increased the frequenting of these establishments.

The first British newspapers were the *Corante* (1621) and the *Weekly Newes* (1622). The first daily paper was the *Daily Courant* (1702) and the first evening paper the *Courier* (1792).

Notebook

Anon, Rome, c.1st century

The arrangement of a text on both sides of a flexible surface, which is cut into two sheets of equal sizes, folded and sewn in the middle, or the notebook, is an innovation which resulted in the expansion of the **book**. Until then texts had been etched on to clay or metal, or on to **rolls** of various materials which were essentially animal skins although copper and papyrus were also used, which could have writing only on one side. This innovation, almost an invention, of the notebook seems to have appeared in Rome in the 1st century; its author is not known. The first known notebook, a fragment of a book in Latin on the Macedonian wars, goes back to this time. It seems likely that the invention of the notebook was facilitated by the extended use of **parchment**, which consisted of the skin of sheep, calves or goats and could be used on both sides rather than just one; this enabled the introduction of the **two-sided page** and disposed of the need for the roller. Parchment was invented around the 2nd century BC in Pergamon in Turkey and by the 1st century its manufacturing technique, particularly at the tanning, polishing and bleaching stages, as well as its stabilization by stretching it on frames, was sophisticated enough for it to be increasingly successful in the Mediterranean regions and therefore also in Rome. The later invention of **vellum**, which was finer since it was made from the skins of young animals, was to increase the success of parchment. It caused the notebook to be gradually adopted, which for the first time allowed books to be carried on voyages and to be read comfortably.

The skins which were prepared of parchment and then of vellum were all more or less the same size. Depending on the size of book desired, the skins were folded twice, **in quarto**, to make four leaves, or three times, **in octavo**, to make eight leaves, and sometimes four times, **in sextodecimo**, to make sixteen leaves, designations which are still used today.

Paper

Anon, Egypt, 3000 BC; Ts'ai Lun (Cai Lun), 105

The ancestor of modern paper was the Egyptian papyrus, the oldest specimens of which date back to approximately thirty centuries before Christ. **Papyrus**, from which the word 'paper' is derived, was made from the fibres of a plant which was very common in the valley of the Nile. This was scraped off, intertwined and pressed by craftsmen, and then left to dry. It was in 105 that Ts'ai Lun (Cai Lun), a member of the Imperial court of Peking, made paper from cellulose scraps for the first time; these consisted notably of mulberry tree fibres, hemp ... and old rags and fishing nets, which were steeped in water, crushed and pressed. The use of paper spread throughout the whole of China at first, then Asia and thence to the rest of the civilized world. By the 14th century there were **paper mills** all over Europe.

Until 1798 paper was made in separate, shaped sheets; that year the first process of continuous manufacture was invented by the Frenchman Nicolas Louis Robert, using the principle of the conveyor belt. In 1807 the Englishmen Henry and Sealy Fourdrinier applied the idea at the industrial level.

Pencil

Anon, c.1300; anon, before 1565; Faber, c.1760; Conté, 1795

At first the use of dry leads which could leave a mark on paper was reserved for drawing purposes; it seems to go back to the beginning of the 14th century in Europe where either **silver or black lead leads** held in a sheath were used. In the 15th century people used finely ground **clayey soils** mixed with **pigments** and then pressed into small sticks for drawing (**red pencil drawings**), which extended the use of dry leads. The first account of the use of black lead inside a **wooden case**, which corresponds to the modern concept of a pencil, was given in 1565 by the Swiss Conrad Gesner, who was not the inventor (this could have been the Englishman William Lee in 1504). Indeed, Gesner does not seem to have been sure of the composition of the lead — it was undoubtedly based on **antimony**. The use of dry leads in drawing nevertheless became widespread during these years, after the opening of a **graphite** mine in Borrowdale, England. As the mine became exhausted, graphite equivalents were sought. It was then that the use of **charcoal** began. This was made out of stems of carbonized charcoal which, being ligneous, withstood pressure satisfactorily. From 1760 the Fabers of Nuremberg commercialized some real pencils, with leads of powdered graphite mixed with a paste made of various substances. The inventor of the modern pencil as we know it, however, was the Frenchman Nicolas Jacques Conté. He made **artificial dry leads** following the process which is still in use, using graphite and fine clays in varying proportions depending on the required result.

The **mechanical manufacture** of pencils was invented around 1822 by the Americans S. Mordan and J.I. Hawkins (ten years after their compatriot William Monroe had set the United States free from importations of European pencils), but it was not patented until 1877, by Alonzo T. Cross.

The invention of the pencil changed the arts considerably, for until then only drawing by pen or by silver point was known. The stroke became more free and flowing.

Photo-engraving

Niepce, 1816

It was in 1816, during research which was to lead to the discovery of photography, that Nicéphore Niepce invented photo-engraving. This invention was based on the properties of **Judaean bitumen**, a substance which has the characteristics of hardening in the light and becoming insoluble in turpentine. By tracing a design on a sheet of glass and placing it on another plate covered with Judaean bitumen, Niepce obtained a **'negative'** of his design, with the parts exposed to the light having hardened and standing out in relief while the others had dissolved in the acid. This invention was to give rise to **heliography**, which was realized in 1875 by the Austrian Karl Kleitsch and which led to the principle of engraving on metal from a photographic negative. The engraved metal plate was covered with ink and then wiped so that the ink remaining in the etched grooves reproduced the document in positive on the paper. Heliography marked the beginning of the age of illustrated printing.

Postal system

Anon, 3rd millennum BC

The first postal systems date back to 2000 years BC and probably began in Egypt; they seem to have consisted of circuits for the distribution of royal and administrative messages by relay. The same practice arose in China a thousand years later. In any case, any centralized power had to have a system of sending orders to the heads of provinces in the shortest possible time. The first **postal relay stations** for which definite proof has been found existed in Persia in the 6th century BC; they were described with great admiration by the Greek historians Herodotus and Xenophon who no doubt deplored the fact that ancient Greece did not have a centralized system, although the state towns each had their own messenger services. The rise of the Roman Empire was inevitably to cause the creation of a vast network for the distribution of administrative documents, still using relay; this was called the **cursus publicus**. The speed of the distribution cannot fail to amaze the 20th-century historian, as the mail covered nearly 250 km in stages, which would mean that a letter sent from London to Cardiff today would take three days ... The mail was taken on horseback of course.

The fall of the Roman Empire did not mean that the postal system also fell apart, because it was the largest and the most regular in the ancient world; indeed, the Romans' successors grasped its advantages far too firmly for it to break down. The Roman postal networks survived until the 9th century, when the growing dilapidation of the road networks — which were also founded by the Romans — as well as the political fragmentation of Europe led to its ineluctable decline. At the same time, the absorption of the ancient oriental provinces of the Roman Empire by the Islamic Empire preserved the *cursus publicus* until the 15th century. All the same, trading requirements pushed the European Guilds of the Middle Ages to reconstitute their own private mail networks and by the 13th century regular networks between the important trading centres of Western Europe were in use, such as between the great commercial towns of Italy, Siena, Florence and Genoa, and the six business centres of eastern France. In the 14th century the importance of the postal system was so widely recognized that, for example, the king of Persia granted the right of free passage to the mail from the Republic of Venice.

Being reserved in theory for state and commercial communications, the postal system was not open to private mail until around 1550, when the prince Thurn und Taxis, who held the monopoly over Imperial mail, asked Charles V for permission to convey the letters of individual people as well; this was effectively the way in which the modern postal service was invented. The ensuing stage was the use of **postage stamps** which transferred the transportation cost to the sender, whereas previously it had been payable by the receiver.

The first postage stamp was printed in Great Britain in 1834 but was not used until four years later. France adopted it in 1848 on the initiative of Etienne Arago, the general manager of the French Post Office.

> The first **prepaid postcard**, was introduced in Austria in 1869 and adopted in the UK the next year — it cost $\frac{1}{2}$d. The first serrated stamp appeared in the UK in 1854; this made it easier to detach single stamps, which until then were cut apart using scissors. The **first airmail** in Britain was run between Hendon and Windsor in 1911 to celebrate the coronation of King George V. The first regular air service was instituted between London and Paris in 1919.

Press agency

Havas, 1835; Reuter, 1851; various, 1861

It was in 1835 that the Frenchman Charles Havas created the first press agency, an organization for the circulation of information to newspapers on the basis of a written agreement. The Havas agency had access to privileged information due to its close contact with government circles. The arrival of the **telegraph** a few years later (see p. 52) enabled the dispatch of information abroad, which meant reduced costs for subscribing newspapers in terms of the up-keep of offices or sending of particular correspondents. In 1851 the German Paul Julius, Baron von Reuter, who two years previously had successfully set up a communications link using **carrier pigeons** between the telegraphic terminal at Aachen in Germany and the one at Verviers on the Franco-Belgian border, had the idea of opening a telegraphic transmission agency in London (he had envisaged setting it up in Paris but the presence of Havas dissuaded him). Then a naturalized Englishman, he limited his activities to the transmission of commercial, notably transatlantic, dispatches; these were then sent out again in Great Britain or on the Continent by carrier pigeon. The Reuter agency only became a press agency in 1858 when it re-transmitted the report of an important speech by Napoleon III.

From 1848 onwards several American newspapers shared the costs of the telegraphic transmission of dispatches. Six of them organized themselves in 1861 to form the Associated Press, whose correspondents kept them informed on the progress of the American Civil War (1861–1865). It was only in 1893, however, that the Associated Press acquired the status of a non-commercial agency.

Printing using movable characters

Pi Cheng (Bi Zheng), 1041; Coster, 1423; Gutenberg, 1439 (?)

The principle of using movable characters for the printing of texts was invented in China between 1041 and 1048 by a man named Pi Cheng (Bi Zheng). These characters were made out of baked clay, a similar method to that used by contemporary typographers, since a whole text composed in this manner was encased in an iron frame. This system was improved from 1297–1298 by Wang Chen (Wang Zhen) who made the letters out of hard wood and used them to write his *Treatise on Agriculture*. He had a catalogue of 60 000 characters in use, which meant that in one month he could print a hundred copies of a local newspaper.

The invention spread throughout vast Asian areas and without doubt arrived in Europe, either because it was described by travellers or because people had Chinese wooden characters as a result of trading, which was very common at the time. In any case, between 1423 and 1437 the Dutchman Laurens Janzsoon, better known under the name of Coster, attempted to adapt it to Latin characters. Since these were appreciably smaller than the Chinese ones, Coster came up against problems which he could not surmount. In the meantime, Dutch and Rhenish craftsmen abandoned wooden characters in favour of metal characters, which were used in the following way: the

It is thought that the Chinese movable characters could have arrived in Europe with the Mongolian troops who invaded Russia in 1240, Poland in 1259 and Hungary in 1283, but there is no formal proof of this.

characters were made in relief in brass or copper and then placed, letter by letter, on to a surface of ductile material such as clay or warm lead in order to reconstitute the text; finally, lead was melted on to the mould made in this way and the new mould was then ready for printing. The process evokes the **casting** in more modern presses and it would have constituted the departure point for printing, apart from the fact that neither the depth of the casts nor the alignment of the characters obtained was satisfactory and each casting tended to dull the original character. **Metallographic printing** was nevertheless practised after a fashion for several years, notably in Strasbourg.

This technical improvement served as a starting point for the German Johannes Gutenberg who had the idea of using movable characters again in the following way: the moulds of the letters were made separately from brass or copper models, in order to be cast as individual lead characters. These were arranged on a wooden plate and then linked together to be used in the printing. Around 1439 Gutenberg also invented the **press**, of which he is the original creator, which enabled him finally to obtain a deep and regular impression of his plates and texts. Around 1475 the German Peter Schöffer improved Gutenberg's technique by making steel moulds. These enabled the casting of characters in copper, which were more precise, more regular and more resistant to wear and tear.

Publicity

Anon, Greece, 3000 BC

The circulation of information of a commercial character is at least about five millennia old. The first evidence for it is a notice offering a reward to anyone able to find an escaped slave; it dates from 3000 BC and was discovered in Thebes. Roman and oriental city streets were full of signs vaunting the goods in the shops. Moreover, public criers were the heralds of spoken publicity for barbers, shoemakers and other professions. The printed press rapidly became a medium for written advertising; in 1652 the British *Mercures* included publicity for cafés. The writer and diarist Samuel Johnson wrote in 1758: 'There are so many advertisements now that they are very carelessly considered and so in order to capture the attention it is becoming necessary to make magnificent promises . . . '. It was only after World War II that the term **advertising** gave way to that of **publicity**.

Publishing

Anon, Rome, c.1st century BC

The **reproduction of books** happened long before the invention of **printing using movable characters** (see p. 48). It was probably carried out during the 3rd or 2nd millennium BC in the Near and Middle East, but it differed from publishing in the modern sense of the term by the fact that the copied works were not sold; the copies, of religious or political texts, were made only to avoid being destroyed by time and they constituted a privilege of the ruling castes.

In all likelihood it was in Rome during the 1st century BC that **commercial publishing** appeared. The authors would hand the original of a text over to a **copyist** and it would be copied several times and sold to whichever customer asked for it (the works were relatively expensive, given that the copies were made on **parchment**, a costly material — see p. 44). The copyist then turned the **rights** to the author, which is much the same idea as in commercial publishing today. After the fall of the Roman Empire the practice of copying passed to the monasteries, but it went back to being non-commercial, or for restricted trading only. It was the creation of **universities** in the 12th century in Europe that restored the publisher's profession. Printing using movable characters obviously gave it a considerable boost.

Modern publishing can be thought to have been created by the British **Copyright Act** of 1709 which protected the rights of an author for 21 years, enabling publishers to republish a text without paying for rights after this time had elapsed, but especially assuring the author a **serial right** and therefore a stable income, which was proportional to the sales.

Serialized novel

Chapman and Hall, Dickens, 1836

The rise of the serialized novel is a major socio-cultural phenomenon. From it the literary works now called 'popular' are derived, including both adventure stories and the great literary socio-historic portraits. The great popularizer of this literary form was Charles Dickens, and the first novel published in this way *The Pickwick Papers*, issued in 20 monthly parts by the publishers Chapman and Hall. They had derived the idea from Dickens's journalistic sketches contributed to the *Monthly Magazine* and other journals. The stories of Pickwick and his friends were so popular that before that serial was completed, *Oliver Twist* had begun to appear at monthly intervals in *Bentley's Miscellaney*. All Dickens's novels were first published in this form, for which they were ideally suited. His characters were more important than the plot, which tended to ramble; a new dimension of suspense was added to the enjoyment of the novel.

Stenography

Xenophon, 5th century BC; Tiron, 63 BC; Bright, 1588

According to most available clues, stenography or 'shorthand' (from the Greek *stenos*, 'narrow', and *graphein*, 'writing') goes back at least to the Greek historian Xenophon who used it to transcribe his conversations with Socrates. In 63 BC Marcus Tullius Tiron, who was freed from the house of Cicero, devised the first codified system of stenography. This spread widely throughout the Roman Empire and must have been easy, for it was used for about ten centuries without significant modification. At that time, stenography was also called 'tironian writing'. Thomas Becket, Archbishop of Canterbury in the 12th century, began to show interest in stenography again, but it was the Englishman Timothy Bright who in 1588 was the first

Some of the first and most illustrious stenographers in history must be mentioned: Cicero himself, the emperor Titus and Julius Caesar.

to recast and adapt tironian stenography to a modern language, English. It was a long, cursive writing. Since then a great many inventors have proposed their systems of stenography; the most famous are the Englishman Isaac Pitman, the Irishman John Robert Gregg, and the Germans Wilhelm Stolze and Ferdinand Schrey in the 19th century, and the American Emma Dearborn around 1925.

Stereotype

Ged, 1725; Hoffmann, 1784; Herhan, Didot, Gatteaux, 1793–1797

Stereotype is a plate-making process which produces a relief printing plate using the intermediate stage of a **cast mould** taken from a body of movable type. The mould with its hollowed-out characters is raised, but the stereotyped page is recessed. This process constituted an essential stage in the development which led to **printing using movable characters** on modern **rotary presses**.

Stereotype was invented by the Rhenish craftsmen printers at the beginning of the 15th century (see p. 48), and for a long time there was no further development because of the defects in the casting. The slow progress of **letterpress** printing meant that it was not fully used until the beginning of the 18th century. The delay was aggravated by frequent accidents, due to the difficulty of handling the mounted pages; they would fall and become disordered, becoming **pied type** (clumps of letters which then had to be composed again); moreover, the composition of works of several hundreds of pages required shops with extensive type resources. In 1725 the Scot William Ged overcame the problem and succeeded in designing the following technique (well before its time); he made metal plates which could be used for printing from pages created from **type moulds** into which metal had been poured. The typographic characters could be reproduced without mishap. This process had an additional advantage of allowing the stereotypes to be kept to be re-used for a reprinting. Ged tried to introduce his technique in London, but the hostility of the English printers ensured that he was doomed to failure. Nearly half a century later, when two Glaswegian compatriots Alexander Tilloch and Andrew Foulis again took up the idea, they were obliged to work in secret to avoid arousing the anger of their colleagues. In 1784 the German Franz Hoffmann endeavoured to establish this technique in Paris and obtained the exclusive rights to his 'new art' of printing; at the end of that year he and his son published a *Journal polytype des sciences et des arts* which received praise from Lavoisier but this time aroused the anger of the Parisian printers. This system was quicker, which offended them. Instead of adopting stereotype they dismissed it by having Hoffmann's licence abolished and getting the technique banned. Nevertheless, the following year Joseph Carez of Toul obtained authorization from the Lord Chancellor to take up stereotype again; he called it **homotype**. The idea spread rather more easily this time since someone had the idea of using it to make **lottery tickets** which were proof against forgery. The technique itself had not been completely perfected, since pouring the hot metal in to the typographic characters did not give very well-defined plates, due to the deformation during cooling. Between 1793 and 1797 Louis Étienne Herhan, Firmin Didot and Nicolas Marine Gatteaux, who ended up forming an association, devised a method of **cold pressing** of the assembled page.

Telegraph

Anon, Greece, 1084 BC; anon, Greece, 300 BC; Lesage, 1774; Volta, 1777; Chappe, 1791; Murray, 1795; Schilling, 1832; Gauss and Weber, 1833; Cooke and Wheatstone, 1836; Morse, 1837

The principle of descriptive communication from a distance is much older than is sometimes thought. It was by means of agreed signs communicated by a battery of fires over a very long distance that in 1084 BC Clytemnestra learnt of the fall of Troy and of the imminent return of her husband Agamemnon whilst she was in her palace in Argos in Greece. Visual communication from a distance using non-alphabetical signs was probably practised previously in Asia, Assyria and Egypt, and it continued for a considerable time, with the American Indians using it during the first colonial conquests of the New World.

The first communication system using alphabetical signs is reported by the Greek historian Polybius. It was invented in the 3rd century BC and defined each of the 24 letters of the Greek alphabet using symbols which were visible from far away. In turn, prisoners in the Middle Ages invented a system which prefigured the morse code, since the code was based on a mental grid of 25 squares arranged in five rows of five, as in the system described by Polybius, but bringing in the notion of time as well. It was in this way that the letter A, first square in the first row, was transmitted by two knocks on a bell with a long interval; the letter B by two sharp knocks followed by a third after an equally long interval; the letter F, by one knock followed by two knocks with a long interval, etc. The first part of each message referred to the place of the letter, and the second from which row it came. Several optical versions of this transmission system seem to have been proposed, and a few of them locally installed.

In 1791 Claude Chappe and his elder brother effected a great step forward in communications by inventing **optical telegraphy** (as well as the actual word, coined from the Greek roots *têle*, 'far', and *graphein*, 'writing'). Optical telegraphy is based on a network of towers built on hills. These are topped with a vertical pole on to which a pivoting T-bar arm is fixed. At the ends of this bar or regulator, the movements of which are controlled by ropes, there are other mobile parts or indicators. It follows that the instrument can adopt numerous

Roman telegraph station *taken from a bas-relief on the Trajan column*

A place in the history of telegraphy should be reserved for the Frenchman Guillaume Amontons, who thought of using a telescope to observe letters fixed to the blades of a mill; this was at the end of the 17th century and Amontons — who was deaf — designed an effective communication system between Meudon and Belleville using this idea. It is generally thought to have been Amontons' idea which inspired Chappe. It was the Convention in 1794 that deserved the credit, one year after Chappe had submitted his idea, for establishing the Paris-Lille telegraph line, along which messages travelled at 200 km/h.

Chappe conducting the experiment of the first aerial telegraph in front of the notables of Paris on 2 March, 1791

positions, 49 in total following the Chappe system, which constitute as many signs of a code. A network of towers was installed in this way and demonstrated the efficacy of Chappe optical telegraphy, which is also known as **semaphore**. One hundred and twenty towers built between Paris and Toulon, for example, enabled a message consisting of 50 signals to be sent in less than an hour, nearly one hundred times faster than the fastest mail service. The system was very successful and was adopted as far afield as India and Egypt. In 1794 the Chappe telegraph had its hour of glory when it announced the French victories over the Austrians at Quesnoy and at Condé-sur-Escaut. It was modified and improved by the Englishman George Murray in 1795, on behalf of the British admiralty, so that it consisted of a notice-board of six pivoting flaps which could make 64 different signs. It was also successful in the United States.

At the same time the principle of **electric telecommunication** was beginning to develop. Proof of evidently fruitless attempts to use electric current to transmit **coded impulses** has been found dating from 1727. However at the time no one knew about the insulation of electric wires, and the strongest currents which could be used were so weak that **amplifiers** were needed at the receiver; at the time no one knew how to make such accessories. In 1753, however, an article appeared in Scotland which created a remarkable stir, and which was signed only with the initials C.M. It proposed to build a communications system consisting of 25 independent electric lines, each corresponding to a different letter of the alphabet; the electric signal sent would give an impulse to a very light body, such as a ball of elderwood. It is not known to this day who the mysterious inventor was, but the idea was applied for the first time by the Swiss Georges Louis Lesage. It can therefore be said, from a strictly historical point of view, that the first electric telegraph was made by Lesage.

In 1777 the Italian Alessandro Volta proposed to use his discovery of the voltaic battery, which would at last ensure a reliable source of current, for the installation of a telegraph line supported by poles between Milan and Como. Chappe, who still had not completed his optical telegraph, thought it a remarkable idea and subsequently undertook to link up two clockwork movements to the two ends of the circuit: he proposed to make the needle point to figures which would designate letters according to a code. Strangely enough, Chappe did not think of a much simpler system, which would have been a dial marked with 25 letters. In any case he became discouraged and went back to his optical telegraph. His idea, which was to be taken up again after various modifications by more than one inventor, was made for the first time in Hammersmith, England, where a Chappe electric telegraph was put into operation. With this model every impulse would make a turning disc with a single window reveal a letter or a figure on a dial inscribed with 35 signs. In 1810 the Englishman Coxe, and the following year the German Samuel Thomas de Soemmering, thought of dispensing with Lesage's elderwood balls and using instead a **voltmeter** which would simply indicate the differences in potential of each of the 25 lines of the Lesage telegraph.

In 1820 André Marie Ampère advocated the use of the **Œrsted effect** which had been discovered the preceding year, and describes the deviation of a magnetized needle under the action of an electric current. It was to be used in such a way that each wire placed above a magnetized needle would make it turn on a scale according to the difference in potential. Great progress was first made in 1825 with the possibility of increasing the intensity of the magnetic fields by including rods of soft iron in the coils, and then in 1829 by the American Joseph Henry with his discovery of the laws governing the design of the coils and the voltages to be applied to obtain the maximal effects. In 1831 he succeeded in making a magnet pivot and sound a bell from a distance of 1.6 km. In 1832 the Russian baron Schilling applied Ampère's first idea and used **magnetized needle telegraph** to link the Tzar's summer palace at St Petersburg to his winter palace. In 1833 the Germans Carl Friedrich Gauss and Wilhelm Eduard Weber built a telegraph consisting of two copper wires between two points, this time situated

2.3 km apart, and on top of which a new system was installed: a **mirror galvanoscope**. The needle of the galvanoscope, which reacted to differences in intensity, was equipped with a small mirror, which when observed from a distance enabled the slightest angular variation of the needle to be detected; this could move by five degrees to the right and the same to the left, and the combinations of variations served as a basis for an alphabetic code.

In 1836 the Englishmen William Fothergill Cooke and Charles Wheatstone returned to the multiple wires. They devised a dial of five magnetized needles, one for each wire (the sixth wire served as a common earth wire for the five others). In 1845 they simplified the dial so that it only had one needle. This system of **needle telegraph** was to be used in Great Britain for giving instructions to tramway drivers, initially on the Paddington–West Drayton line.

In 1837 the German Carl August Steinheil, whose name has gone down in posterity because of inventions in **photometry** and **optics**, thought of devising a way to transcribe the signs corresponding to the letters on to a roll of paper, and thus invented the first **teleprinter**. During experimentation he discovered that the ground is a conductor and this meant that an **earth wire** was not needed. The discovery of electric transmission by a single wire would considerably simplify telegraph construction. During the same year, the American Samuel F. Morse studied his compatriot Henry's discovery in greater depth and invented a reliable **electromagnet**, which he patented. He also invented the **morse code**, after having toyed briefly with the idea of a system of combining the typographic signals. With due respect to Morse, it could be said that he actually only invented the code which was named after him, since the electromagnet proceeded from Henry's discovery and single wire transmission was the work of Steinheil. However, it was Morse, out of all those who had worked on the telegraph until then apart from Chappe, who knew best how to exploit his patents. In 1843 he had the **first telegraph line** built over a distance of 60 km between Baltimore and Washington and thenceforth his telegraph networks proliferated, giving rise in 1856 to the Western Union Telegraph Company, which still exists. It must be noted nevertheless that morse was the simplest system of all, since the code consisted of only two types of sign, a short and a long, which could be combined ad infinitum; furthermore, the transmission and reception equipment was of a very simple design.

In 1845 the **first underwater cable** in the world was laid between New York and Fort Lee, and in 1851 the first international connection took place between England and France, using a submerged cable which was insulated with gutta-percha. It was only a year later that France finally adopted the electric telegraph, and again a needle system was preferred, the Foy-Bréguet, which was really less practical than the morse system. The telegraph had just brought about a great chapter in communications. The ensuing one was to be that of wireless telegraphy, the beginnings of which overlap with those of the radio.

Since Michael Faraday's discovery of **electromagnetic induction** in 1831, several scientists foresaw the possibility of communication by electromagnetic waves, without the use of a conductive wire between two distinct points. The concepts concerning these waves were vague; the Scot James Clerk Maxwell clarified and organized them in 1865 according to his theory on electromagnetic waves, postulating that there are many different frequencies of electromagnetic waves. This was but one theory, which took a long time to be accepted at a point when people were already very partial to rational explanations. It was verified only in 1885, when the German Heinrich Hertz built a very simple small device which consisted of two copper balls fixed to rods a few millimetres apart. These were charged with electricity so that a spark jumped between them; this was proof that 'something' really did move and that this something had every chance of being electromagnetic waves. Two years later Hertz carried out the experiment again with something he called a **spark gap**, but this time he installed a brass wire hoop a few metres away. There was a gap of a few tenths of a millimetre in the hoop, where Hertz maintained that he received the sparks produced by the spark gap. This can be said to be both the verification of Maxwell's theory and the first connection by **Hertzian waves**. One wave emitted at one point had been received at another. It was the Italian Guglielmo Marconi who followed up these experiments most systematically. His efforts were made considerably easier by three earlier workers: Edouard Branly, who invented the **iron filings tube** or **coherer** in 1890, which allowed the electromagnetic waves to be captured in a much more regular way than the hoop resonator achieved; the Englishman Oliver Lodge, who had designed the first wireless telecommunication device; and the Russian Alexandre Stepanovich Popov, who had successfully conducted experiments in Cronstadt harbour using wireless telegraphy between the ships of the Imperial fleet and the land. In 1898 Marconi made a cross-Channel connection between Wimereux and Dover and sent signals across the Atlantic in 1901. The invention of telegraphy appeared therefore not as a stroke of genius by a Morse or a Marconi, but as a series of modest stages spread over at least a century, each one being provided by an inventor supported by his predecessors. This is one of the most characteristic inventions in history.

Typewriter

Burt, 1829

It is rather difficult to establish the originator of the typewriter. Prototypes of it, either completed or still at the project stage, seem to have existed since 1710. In that year a portable machine may have been made by a man named Hermant which 'enabled fast writing, in printed letters'. It was discovered in 1728 by the abbot of Olivet from the French Academy who then told Bouhier the presiding judge of the Dijon Parliament; the machine certainly did exist and it seems to have been satisfactory, but as we do not have a copy of it the paternity must remain in suspense. In 1714 in England Henry Mill was granted a patent by Queen Anne for a similar machine, but no example of it is recorded.

In 1829 a patent was granted to William Burt of Detroit, USA, for a machine called a **typographer**. The type was mounted on a rotating semicircular frame, and depressed against the paper by the operation of a lever. Attempts were made by many other hopeful inventors, but the results from the machines were often slower than writing by hand. Christopher Latham Sholes began work to make a practical version in 1867, and after making improvements was able to sell his machine, the 'literary piano', to a firm of gunsmiths called E. Remington and Sons. The Remington machine was sold commercially from 1874.

The firm Remington made 25 000 of its Model 1 but sold only 1200 of them. The instrument was very expensive ($125, or roughly £70 at the time), very heavy and difficult to handle. Doctors accused it of causing consumption and bringing on schizophrenia and Queen Victoria became angry when someone presented her with a '**chirographed**' document, for this was the term for 'typed' at the time. For a long time the writer was unable to see the text as he typed. The first machine to produce a visible text was made in 1890 by the American John N. Williams and it was also the one which made the machine popular.

The first writer ever to submit a typed manuscript to his publisher was the American Mark Twain who had bought himself a Remington and made great use of it.

The first **electric typewriter** was made in 1872 by the American Thomas Edison and the **teleprinters** in press agencies were the first to put it to use.

Daily Life

Even though clothes had changed from antiquity to the end of the 19th century, and had become increasingly complex, very few aspects of daily life were actually improved. In the area of hygiene, for example, in most European cities, running water was a rarity, whereas most Roman homes had enjoyed this facility. The hygiene of Europeans was detestable compared with that of the citizens of the Roman Empire, and when mains drainage was finally installed there was a general outcry. 'Orthodox' people proclaimed that it was an attack on private life.

When matches were invented, they were dangerous to use and their stench was offensive. Daily life in Europe and America benefited from only one significant step forward at the beginning of the century: street lighting, which was firstly by means of gas, and then arc-lights. However, it was the industrious inventor of the electric light bulb, the American Thomas Alva Edison, who was to chase away the darkness once and for all from both houses and towns.

It was not until after World War I in the 20th century that the homes in Western towns were to have the standards of cleanliness and comfort which Roman homes had offered twenty centuries before.

Romans not only knew how to build houses of several storeys, but also invented the lift and air-conditioning. Locks in Cicero's time could only be picked by expert thieves, and central heating by means of hot-air pipes meant that the Romans did not get cold in their houses when the temperature dropped.

Air-conditioning

Anon, Greece, c. 350 BC

There are two aspects to changing the temperature inside buildings, **heating** and **cooling**. Heating by methods other than an **open fire** was, if not invented by, then at least practised by the Lacedaemoniens in Greece around 350 BC. It consisted of directing air which had been heated by an **underground fire** into baked clay pipes under the floor. This system was improved by the Romans who designed real heating systems which extended through the walls of the houses, right up to the upper floors. After falling into disuse in the Middle Ages, this method of heating seems to have been taken up in Greenland in the 13th century; Greenland was then a colony of the Norwegian crown. In times of prosperity the towns of Greenland were heated in this way, using hot-water springs — which even allowed the cultivation of orange trees under glass. The Roman method of central heating was forgotten in the rest of the world; it was only re-invented in Europe at the end of the 18th century.

Cooling seems to have been practised for the first time around 1500 in the château of the marchioness Isabella d'Este, based on a principle by Leonardo da Vinci: a **paddle-wheel** plunging into a covered basin of water fed a very large set of **bellows** which injected this air into the marchioness' apartments; it does not seem to have been copied.

The first patented cooling system was invented in 1831 by the Englishman Jacob Perkins. It was first installed in a building in 1837, in the London Parliament which was in fact the first building in the world to be cooled. The principle was as follows: fires installed in the attics created a flow of air in the piping which took in cool air through underground openings situated above the Thames. The cool air was distributed throughout the building via secondary canalizations.

The cooling system patented by Perkins was known and used for centuries in underground mines, at the openings of which braziers were kept going to create a draught and thus clear out the warm air. Mention must also be made of the very ancient system practised in India which consisted of moistening mats and hanging them at the windows.

Bank-note

Anon, China, 8th–9th century

The **registered promissory note** for payments from a distance is a very old custom and difficult to date, but traces of it are certainly found from Imperial Rome. It was towards the end of the 8th century or at the very beginning of the 9th century that the **unregistered promissory note** appeared in China, which was equivalent to the modern 'cash voucher'. It was 'nationalized' in 812 and used for the payment of taxes. In the 10th century in the province of Szechwan there were no less than 16 private banks which were authorized to produce notes such as these. The

The first coloured bank-notes (in six colours) appeared in China in 1107. It was only in 1661 that a European bank adopted the Chinese custom; this was the Riksbank in Stockholm.

In 1126 the imprudent issue of 70 million bank-notes which were not covered by a guarantee in gold caused catastrophic inflation in China.

system was nationalized again in 1023, however, and these banks were replaced by a central bank which was the first to produce notes of various denominations. The first bank-notes were valid for only three years.

Forgers were quick to manufacture false bank-notes. It is not known when the first false note was seized, but it is a fact that in the year 1697 alone, 300 people were hanged in the city of London for the crime of counterfeiting, which at the time was punishable by death.

Chess

Anon, China, 4th century BC

Few inventions have been the object of as many attributions as the game of chess. The Babylonians, Jews, Egyptians, Scythians, Persians, Chinese, Hindus, Arabs, Greeks, Romans, Castillians, Irish, Araucanians, Gauls, Japhet, a Singhalese king, Rava, Xerxes, Semiramis, Queen Zenobie and the Brahman Sissa were all credited with it in succession. The most generally accepted hypothesis is the one defended by the English historian H.J.R. Murray in his *History of Chess* published in 1913: this was that chess was invented in India in the 7th century and from there it was imported into Persia and, during the ensuing centuries, to the rest of the world. Its Hindu name '**chaturanga**' means 'the four army bodies', elephants, horses, chariots and soldiers; it was a game for four people where the right to play was decided by dice. This theory, which seems quite plausible, is maintained elsewhere by numerous studies. There is no doubt that it was in India, and more specifically in Hindustan, that chess first developed considerably. However, the research by the sinologist Joseph Needham must be acknowledged, according to which a game existed in China which was played on a board using bronze pawns which were engraved to differentiate their roles; according to the historian Mencius this game, called '**i**', dates from the 4th century BC. It was a version of a game of astrological inspiration, the first version of which would even have existed in the 6th century BC. Still according to Needham, the figurines of baked earth found in a tomb, dating from the 2nd century, were played in a game called '**liu-po**', that is 'the six scholars', which existed in the 3rd century BC. It follows therefore that the origins of the game of chess were Chinese and predated the Hindu version, which was derived from it, but from which the later Chinese versions were themselves derived.

The date when chess was introduced into Europe is not known, but it seems to have been before the first crusade. This is attested by a letter from Saint Pierre Damien to Pope Alexander II, dated 1061, in which the saint, who was then the cardinal bishop of Ostia, stated that he had imposed a punishment on a bishop whom he had found playing this game. Moreover it is known that the Byzantine emperor Alexis Comnène, who died in 1118, relaxed by playing chess, as is reported in a letter by his daughter Anna.

It was in France around the 15th century that the rules of the game of **chaturanga**, the form practised by the Europeans since its introduction, evolved towards those of the contemporary game of chess.

Cigar and cigarette

Anon, pre-Colombian America, unspecified time; anon, Spain, after 1518

The tobacco plant, its cultivation and consumption seem to be exclusively American. It is not known exactly when the growing and use of it began in pre-Colombian America, but with good reason the word 'cigar', of which 'cigarette' is the diminutive, is assumed to be derived from the Maya '**sik'ar**', which refers to tobacco leaves rolled in maize leaves.

It was in fact the pre-Colombian Indians who invented the treatment and fermentation of tobacco leaves, which was intended to make them less acrid. The tobacco industry was transplanted by the Spanish conquistadors to Haiti, and then to Europe in 1530, but in 1518 small cigars called '**cigaritos**' were exported to Europe by the Spaniards. The cigar and the cigarette had very different destinies; the cigar was very popular in America, where it was mass-produced in 1785, whereas the cigarette did not appear in the New World until 1860. In contrast, the cigarette was immediately successful in the Old World where it came into widespread use in Turkey and China.

Cigars and cigarettes were rolled by hand at first. **Cigar-rolling machines** did not appear until 1919 in the United States, whereas **cigarette-rolling machines** were patented in 1880 by the American James Bonsack.

> Until 1920 the world consumption of cigars was equal to that of cigarettes; since 1950 it has been four times less.

Drinking water (distribution of)

Anon, Mohenjo-Daro, 3rd millennium BC; anon, China, 500 BC; Simpson, 1829; anon, after 1850

Supplying the public with drinking water has assumed two aspects over the ages: until the Industrial Revolution it was the **supply** itself which was important, and after the beginning of the 20th century, both the supply and the **suitability** of the water for drinking were considered. The first inventions in this domain were very old and anonymous: the civilizations of the Indus valley and in particular those of Mohenjo-Daro in the middle of the 3rd millennium BC practised the digging of **wells** — undoubtedly the oldest known — and ensured they were **hygienic** by lining them with baked bricks. Some two millennia later the Greeks installed networks for water distribution built out of stone and baked earth, the former for the **reservoirs**, the latter for the **pipes**. This too was an invention to the credit of Greek engineering, even if it was the Romans who developed the technique. Also to be mentioned to the credit of the Greeks is the invention of the first known system of distribution by **atmospheric pressure** which was found in Pergamon dating from the 2nd century BC. An **aqueduct** conveyed the water at a height of about 120 m. From there the water was sent through a network of baked earth pipes — which were sometimes varnished on the inside — towards public fountains, water tanks, baths and private houses. In 312 BC the Romans, who were very keen on water and baths — they used some 200 l per person every day, which is approximately the same as the water consumption in contemporary industrial civilizations and which was unparallelled for centuries — had built large aqueducts for bringing the water from nearby mountain springs to Rome. In 305, six centuries later, the city of Rome alone was served by 14 aqueducts. The rest of the Empire also benefited from similar structures, the repercussions of which extended to the whole of civil engineering. The fall of the Empire sounded the death knell for work of this kind. As well as many other woes, the Middle Ages also suffered from problems with the supply of drinking water. The people provided themselves with water when and where they could, particularly from the rivers, which also served as sewers. The two critical aspects of the supplying of urban centres with water became very evident: the need for a regular and sufficient supply and the suitability of the water for drinking.

The former need was the most obvious, but it was not until 1582 that a decision was made to make provision for it: this was the year when the Dutchman Peter Morice installed the first **hydraulic turbine** in London to provide the town with water from the Thames. It was a productive example, for at the beginning of the 18th century an aqueduct was even built in Mexico. The first town since the Greeks to equip itself with a water distribution system using pipes was again London, where in 1619 a private company built a system for supplying private houses with water. In 1761, the **advent of steam** enabled this network to be enlarged, not only in London but in many other towns.

The combination of **industrial expansion**, the **growth of urban populations** and the consumption of river-water meant that the people were drinking polluted water. It took a long time for this to be remedied, because the **role of germs** had not yet been discovered and the doctors who claimed that disease spread as a result of 'miasmas' disagreed, often violently, with those who maintained that they were transmitted by contagious physical agents. Here the credit goes to the Scot James Simpson who in 1829 installed the first known **water purification system**; for this Simpson used and re-invented the method of **slow filtration on sand-beds**. His aim was only to provide relatively limpid water, but in 1854 the Englishman John Snow proved that the cholera epidemic which ravaged the capital could be traced back to a contaminated well. The following year the city of London, which at the time was the most advanced in the world with respect to hygiene, imposed filtration as a precondition for all river-water. After Pasteur's formal demonstration of the **role of microbes in causing diseases**, active methods of water sterilization were sought, including **chlorination**, **nitrification-denitrification**, **ozonization**, and **filtration through active carbon**. It was chlorination which had the best results, because of its simplicity. Ozonization was discovered in 1783 by the Dutchman Martinus Van Marum but was not adopted until the 20th century. Active filtration only became widespread later, notably after a cholera epidemic had broken out in Hamburg in 1892, which had taken place even though the town used passive filtration stations. These stations turned out to be incapable of decontaminating the water of the Elbe because it was too heavily polluted. Until 1850 only 16 American towns had municipal water tanks for the supply of drinking water. It was only during the following half century that the distribution system expanded.

Since ancient times, therefore, the distribution of drinking water, essential to the life of society, had benefited from only one real invention: Simpson's passive filtration. The rest of its evolution was dictated by the work of Pasteur and the adaptation of purification methods.

Eau de Cologne

Anon, Montpellier, 19th century

The first known mention of a non-oily scented liquid was that of **Angel water** which was the speciality of Montpellier from the 19th century. It was used for cleansing.

Thereafter, several other waters came into vogue, breaking with the ancient tradition of oily perfumes. The first one, which was composed of alcohol and citrus and known as 'eau de Cologne', made its official appearance in France in 1855, presented by a member of the Farina family of perfumers.

Hot water heating

Cruges and Price, 1829; Perkins, c. 1830

The development of a sense of **comfort**, which went hand in hand with the rise in the quality of life and the increasing number of multistorey **residential properties**, both of which followed the Industrial Revolution, as well as the ensuing **urbanization**, quickly aroused new requirements concerning **air-conditioning** and **central heating** in particular. Neither **fireplaces**, the **draught** of which had been improved since the 18th century, nor fixed or portable **stoves** sufficed for the new comfort requirements. The widespread use of **coal** from the end of the 18th century and the mastery of the construction of **boilers** led naturally in 1810 to buildings being heated by **steam**. This consisted of an adaptation (not an invention) which was relatively simple to work, since the sizes of the fire and the boiler simply had to be adapted to the networks of piping which conveyed the steam and which were evidently installed in closed circuits, with the return taking place in the direction of the boiler.

The first invention in the area of heating consisted of replacing the steam with the **circulation of hot water**. Paradoxically, this invention came long before the patents to protect it which were submitted in 1829 by the Cruges brothers, who were English, and their compatriot Charles Fox Price, since the first buildings heated using hot water were in France in 1777 and remained so until 1845. The water circulated by **atmospheric pressure**. Around 1840 the Frenchman Léon Duvoir had the idea of going back to steam, but this time injecting it under pressure; it was under a pressure of 5 atmospheres that he diffused the steam in the pipe circuits which, for the first time, heated the High Chamber in Paris. He thus made a significant economy in the amount of heat used, which the Englishman Jacob Perkins increased again by raising the pressure to 15 and 20 atmospheres.

It was Perkins who around 1830 invented the **radiators** of modern central heating, which were in fact coils and were placed horizontally at first.

The clearing of the rubble of the houses in London after the bombing of World War II uncovered the remains of Roman houses dating to the occupation of Britain in the 1st century BC. Consequently the archaeologists learned that the Romans had used central heating too, which had consisted of an installation of underground fires heating up brick plates which were incorporated into the walls; they were therefore heated in the first place by radiation. This system was used for a long time and was modified only at the beginning of the Middle Ages, when fortified towns were built, followed by fortified castles, which hardly lent themselves to the adoption of heating. Open fires were used then. The first fireplaces date only from the 14th century.

Insurance

Anon, Babylon, 4000 BC; anon, India, 600 BC; anon, Greece, 300 BC; anon, Italy, 15th century; anon, England, after 1666; Franklin, 1752; Lloyd's, 1779

The practice of insurance is extremely old, for a form of it has been found which dates from Babylon in 4000 BC. It consisted of borrowing funds for a commercial venture, funds which were deposited with a third party, who acted as banker. The smooth running of the business and the good faith of the operator were thus guaranteed by these funds which he totally or partially owed. Insurance therefore began as a **guarantee**. The practice was also found in India in 600 BC, but these examples are mentioned only as landmarks, for this type of insurance was practised in many other countries, such as China and Egypt, and in Rome of course.

This type of security developed towards the modern conception of insurance after the Great Fire of London in 1666; it was actually after this disaster that **fire insurance** appeared, as did the first specific insurance companies. The principle behind it was specifically changed too: the annual fee of the cover, which was proportional to the value of the house, was indefinitely renewable, and the deposit was no longer made with a third party but guaranteed by a bank. In 1711 two large insurance companies in the modern sense appeared, the London Insurance Corporation and the Royal Exchange Insurance Corporation, which were extended to deal with other material goods as well as houses. In 1752 Benjamin Franklin founded the first American insurance company, the Philadelphia Contributionship, and in 1759 the first **life assurance** company was founded, also in the United States, which was called the Presbyterian Ministers' Fund.

In the meantime, insurance had evolved under the influence of a London café-owner called Edward Lloyd, whose establishment in Tower Street had been used as a meeting place for the people in the profession since 1688. Contrary to what one might assume, Lloyd, who died in 1713, never founded an insurance company. Paradoxically, he kept his café and was content to publish a bulletin of legal and maritime news, *Lloyd's News*, which subsequently became *Lloyd's List*, giving the arrival and departure dates of ships as well as maritime and professional information. The café survived him as a meeting place for the men in the profession and in 1774 it was set up in the Royal Exchange buildings, remaining a sort of club by **tacit agreement** and **mutual consent**, without status, according to a typically British pattern. Its influence was considerable, however, since Lloyd's acted as adviser to the Admiralty. In 1779, and as always without legal status, Lloyd's introduced the **printed contract**. The company only gained its present legal status in 1871, in order to have a unified control over its members. Lloyd's was also the first company to practise **re-insurance**, which involves guaranteeing other companies' contracts.

The beginnings of modern insurance were sometimes rather turbulent, particularly after the South Sea Bubble scandal in Britain in 1720, after which financial con men were dealt with ruthlessly, and after the Chicago fire in 1871 as well as the earthquake and fire in San Francisco in 1906, when numerous companies were ruined. Eighty-one companies went bankrupt between 1870 and 1877.

Iron

Anon, China, 4th century BC; Steely, 1882

When the iron was first used in China, it was an instrument very similar in shape to our modern irons, and was heated by being filled with glowing embers. Traces of it have been found which date back to the 4th century BC. It was apparently only around the 17th century that irons using fire or embers appeared in Europe. This is thought to have been linked to the expansion of the cultivation of the potato, as it had been learned that the starch could be extracted for starching. The electric iron was invented in 1882 by the American Henry W. Steely but its use only expanded with the extension of the Edison company networks.

Lift

Vitruvius, 1st century; Otis, 1851; Edoux, 1867; Siemens and Halske, 1887

The principle of a raised platform inside a vertical cage for the transportation of people or heavy materials was first described by the Roman architect Vitruvius. The elevation was carried out using a **counterweight**, the raising and lowering of which was controlled by a **crank pulley** outside the platform. It seems likely that lifts of this type could have been used in multistorey Roman homes, where they would have been operated by slaves. Mechanisms of this type called '**flying chairs**' appeared during the 17th century at the court of Savoie, in Chantilly, in Paris, and Louis XV is known to have had them installed in Versailles to take him up to the Dauphines' rooms. In the flying chair designed by Villayer one could control the crank of the pulley oneself.

The name of the Englishman who in 1800 thought of using a **steam engine** to power the lifts is not known; this motor was installed on the roof and controlled the winding of the cable (which was actually a strong hemp rope) on to a drum. In 1851 the American Elisha Graves Otis invented a **safety system** which prevented the rope from waving about by securing it into a rail and blocking it with a series of claws. The first passenger lift was installed in 1857 in the Haughout stores in New York. In 1867 the Frenchman Léon Francois Edoux invented both the French word for 'lift', which is 'ascenseur', and the **hydraulic column lift** which he presented at the Universal Exhibition that year. The device functioned by the injection of water under pressure into a vertical shaft where the column rose and fell like a piston in a cylinder; the injection of water was achieved using a steam engine. When the water was injected, the column rose; when it drained away, the column descended. In 1889 Edoux built a 160 m-high lift for the Eiffel tower. These lifts were twenty times faster than their predecessors which worked on traction. In 1887 electric energy was applied to the traction of lifts for the first time by the firm Siemens and Halske, for the Industrial Exhibition in Mannheim that same year. The electric energy was applied directly to the rotation of a drum (underground) around which the cable activating the lift was wound. The use of electricity allowed **switch controls** to be introduced in 1894. The following year, an English engineer had the idea of applying the electric energy to the pulley operating the cable, between the lift and its counterweight.

Lock

Anon, Nineveh, c.4000 BC; Barron, 1778; Bramah, 1784; Chubb, 1818; Yale, 1860

The oldest known lock was found in Nineveh in Mesopotamia and its age is estimated to be about 6000 years. It is of the type called Egyptian, because it was especially in Egypt that they were first used, although some have been found in India, Zanzibar, Norway, in the Faroe Islands and in Japan. It consisted of two massive pieces of wood which were shaped so that they fitted into each other, with one piece being 'male' and the other 'female'. These pieces were held together by free **pins**, which passed through the top of the female and the male parts, and which were arranged in a variable order. The key looked like a toothbrush and was a slightly bent shaft which at one end had teeth corresponding to the arrangement of the pegs. When it was inserted below them, an up-and-down movement served to raise the pegs in such a way that they freed the male part. It was a reliable principle, for it was to be found again in Yale locks at the end of the 19th century.

The Greeks used another type of lock which was more primitive, consisting of a stem which had a sort of sickle fixed to the end. Once the key was engaged in the opening of the lock, it made the **bolt** move and pushed it back. These were the first metal locks. The Romans improved them by inventing both the shaped **key-ways** which prevented a foreign key from opening the lock, and the individual cut of the key itself, corresponding to the shape of the key-way. This system enabled them to reduce the dimensions of the locks, for which the keys became so small that they could be worn disguised as rings; it was also adapted to flying locks or **padlocks** which could be used to bolt together two hooks which were superimposed or side by side. The Roman lock was used for centuries, the only later improvements being to do with the cutting of the key-way and the key.

This type of lock could be forced by a kind of brush with metal bristles which adapted to any shape of keyhole. The first step towards greater security in a lock was taken by the Englishman Robert Barron, who in 1778 designed and patented some **levers** which corresponded to the shape of the key and which were added to the **key-ways**. In 1818 his compatriot Jeremiah Chubb improved this type of lock by equipping it with a spring which ensured the lowering of the levers and seriously complicated the robbers' task. This part was called the **informer**.

Meanwhile a third Englishman, Joseph Bramah, had invented another, remarkably clever type of lock. The key of this lock was bored, or hollow, with grooves of unequal lengths; each one of these pushed back blades of different lengths in the lock, which were inserted into it. It was only the exact insertion of the blades in the corresponding grooves which enabled the key to be turned to engage the bolt. His safety lock remained pickproof for 60 years.

The progressive urbanization of the industrial countries and the increase in crime stimulated the demand for locks. Indeed, there were more and more doors, and therefore more and more temptations for the burglars; and so the locksmiths did not leave the situation as it stood. In passing, an improvement designed during the 1850s by the American Robert Newell should be mentioned: the entire lock, which was circular, would turn only when the key was engaged inside, which prevented the burglar and picklocks from examining the mechanism.

The great invention of the end of the 19th century, however, was incontestably that of

> Bramah was so sure of the inviolability of his make of lock that he offered the reward of £200, which was a considerable sum at the time, to anyone who could succeed in picking it. No one was able to claim the prize until 1851, when an American locksmith, A.A. Hobbs was successful.

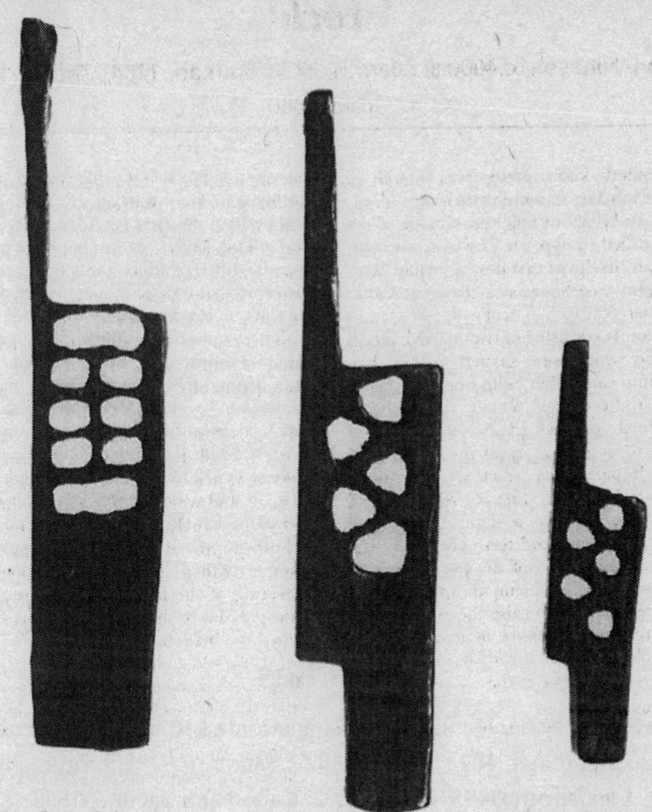

Bolts from Roman locks

Linus Yale, Jr; the indentations of the shaft extended the whole length of the key, activating five pins on springs which only allowed the **barrel** of the lock entrance to turn to engage the bolt if they had been raised at exactly the predetermined heights. This formula which was only a version of the Egyptian lock with the addition of springs to the pins and the application of technological advancement, was advantageous in one particular way: it could be made in an infinite number of combinations.

Up to the beginning of the 20th century only one improvement was made: this was the addition of a **timing system** which allowed the locks of safes in banks to be opened only at a specific time.

Lottery

Anon (Augustus or Nero?), Rome, c.1st century BC; anon, Italy, c. 12th century; François I, 1520

The allocation of a prize as a stake in a game of chance is undoubtedly as old as society itself. However, the chance allocation of a prize in a game of chance in which an indefinite number of people take part seems to go back only to around the 1st century BC; this was the origin of the lottery as we know it today. It was practised and perhaps invented by Augustus or Nero and offered slaves, boats or houses as prizes, but it differed from the contemporary lottery in that it cost nothing to participate; this practice was part of the quasi-institution summarized under the name of '*panem et circenses*', or 'bread and games'. In Italy around the 12th century the right to participate began to be charged for and the practice spread to France, Germany and Austria, where it served to enrich the public coffers. In 1520 Francois I promulgated an edict to make the **blanks** legal, this being the name for the lottery tickets and even the game, which was instituted again in 1539 as a system for enriching the state.

When it was introduced in England in 1569 the lottery was amazingly successful, so much so that it was a lottery which financed the Crown expedition to Virginia in 1612. In 1823 the House of Commons was obliged to apply strict regulations to it because of frauds. In 1890 the American company Louisiana Lottery had become so corrupt that President Harrison had to expose it publicly, and it was made illegal in 1892.

Matches

Boyle, 1680; anon, France, 1780; Cagniard de Latour, 1810; Derosne, 1816; Döbereiner, 1823; Walker, 1827; Sauria, 1831; Schrötter, 1845; Lundstrom, 1855; Pusey, 1892; Cahen and Savène, 1898

The history of matches is strangely long for what seems to be a simple invention, for it lasted for nearly two centuries. As with many others it began with alchemical research, which in this case led the German Hennig Brand to a mysterious thick liquid, which he hoped would turn into gold when it boiled. Brand had just discovered **phosphorus**. The first 'father' of matches was the Englishman Robert Boyle, who discovered in 1680 that thick paper covered with phosphorus would ignite twigs covered in sulphur when they were rubbed on to it. He thought of making matches in this way, an invention which would have been very welcome, for until then the only way of having fire in a house was if embers which could be used to ignite the wood were kept going permanently. Phosphorus was very expensive, however, and the idea was not developed. It is not known who revived the idea in 1780 in France in a form which was as complicated as it was unprecedented. In that year '**phosphorus candles**' appeared in Paris; these were sealed glass tubes inside

The often considerable difficulty in making fire meant that from the beginnings of humanity until the commercialization of harmless matches (which was not until two years before the end of the 19th century), the need to keep a fire going continuously in the home became vested with what was first a sacred and then a moral character.

which there was a piece of thick paper soaked in wax, with its end covered with phosphorus. Since the pure phosphorus ignited immediately on contact with the air, all that had to be done for a flame to be obtained was to break the top of the tube, which was in the shape of a bulb (this second true precursor of the match was also called an '**ethereal light**').

An invention derived from this appeared in Italy in 1785 under the name of **pocket light**: it was a flask filled with partially oxidized phosphorus, in which sticks of wood which had one end coated with sulphur were dipped; thus it was actually only a repeat of Boyle's invention. This was the idea which was taken up and modified slightly by the Frenchman Jérôme Cagniard de Latour in 1810, and which was inverted by the Frenchman Francois Derosne in 1816 when he coated the sticks with partially oxidized phosphorus and rubbed them against a tube coated in sulphur. All these inventions gave rise to other versions and were quite successful. In the United States, for example, for two dollars, (a high price at the time) a small bottle could be bought which contained 50 sticks, each coated at one end with potassium chlorate, sugar and gum arabic; these were ignited by dipping into sulphuric acid! Even more disconcerting was the **resinous match** which was ignited by aiming a jet of hydrogen on to it! The prize, however, goes to another '**portable light**', (in fact all these inventions were meant to be portable): the hydrogen was generated by the action of dilute sulphuric acid on zinc; this gas was then directed on to a platinum sponge in contact with the air. All these objects were bulky. The last one was the work of the famous German chemist Johann Döbereiner in 1823. These technical eccentricities, which were sometimes dangerous, kept on multiplying until a match which would ignite simply by friction was eventually found. Hence the incredible '**Promethean match**', which consisted of a bulb filled with sulphuric acid and coated on the outside with potassium chlorate, sugar and gum arabic, the whole thing being enveloped in paper. The bulb had to be broken with the teeth to ignite the paper. Presumably more than one moustache was singed. The list of surprising inventions, such as revolvers with inflammable capsules, self-lighting cigars and 'Prussian' guns, is too long to be detailed here; the examples cited attest to the ingenious efforts which were made for decades in order at last to find some practical matches.

Steps in the right direction were taken in 1827 when an English pharmacist, John Walker, made '100 sulphuretted peroxide strikables' on behalf of an individual customer, who could in fact have been the real inventor. They were composed in the following way: sticks of nearly a metre in length, the end of which was coated in antimony sulphide, potassium chlorate, gum arabic and starch, were ignited by being pulled sharply between two sheets of glass paper. These were obviously not matches as we know them today, not only because of their size, but also because of the conditions under which they ignited. They generated a series of explosions and an appalling smell, so much so that another manufacturer, who took up Walker's idea several years later, advised people with weak lungs to abstain from using the matches which were given just as evocative a name as the ethereal and Promethean ones before them: they were simply called **lucifers**! Finally, however, really portable matches came into use.

By replacing the antimony sulphide with phosphorus, which had become easier to obtain, the Frenchman Charles Sauria, in 1831, made great progress: the combustion was much more regular, although just as lively and not entirely free from danger, and the smell was less offensive. Sauria did not put down a patent, however, and he lost the rewards of his invention to the American Alonzo Dwight Philips who patented it himself in the United States in 1836.

The danger presented by the Sauria type of matches lay in the phosphorus fumes which the users would inhale and which could penetrate the body through decayed teeth. These gases caused a necrosis of the jaw and they are thought to have killed thousands of workers who were employed in their manufacture, not to mention the children who died because they picked up matches thinking them to be sticks of barley sugar. At the point when the invention seemed to have been completed, it was categorically rejected by public opinion. As well as its organic risks, it actually presented an additional danger: the matches made of white phosphorus tended to ignite spontaneously. In 1845 the German Anton von Schrötter discovered **amorphous phosphorus** and in 1855 the Swede J.E. Lundstrom applied his discovery to the manufacture of the first **'safety' matches** which would only ignite if they were struck against a strip also coated with amorphous phosphorus which was on one side of the box. In 1892 the American lawyer Joshua Pusey invented the cardboard match, which in the same year he began to manufacture in the form known today, in small packets of twenty. These were slow to be successful, for red phosphorus was still suspected of being harmful. It was only in 1898 that the Frenchmen E.D. Cahen and H. Savène managed to make matches that really were harmless, which they did by substituting non-toxic **phosphorus sulphide** for the phosphorus; these were to have the lasting economic success which continues today.

It is interesting to note that the modern, so-called safety matches are not made of phosphorus sulphide, but of antimony sulphide like Walker's matches, with potassium chlorate as an oxidizing agent, and either sulphur or carbon. The machines used for their manufacture, which were first made in 1854 by the American Samuel Gates, Jr, and improved in 1864 by the Swede A. Lagerman for continuous production, have not changed fundamentally since then.

Oil-painting

H. and J. Van Eyck, c.1420

A doubtful tradition would have it said that it was the brothers Hubert and Jan Van Eyck who invented oil-painting around the year 1420. However, even though it is fact that Jan Van Eyck did carry this new technique to a high point of perfection (until then paintings had been made by the **tempera** method, that is, using **egg-white**), it is also true that the use of oil as a medium for coloured pigments goes back at least to the 11th century, if not further. It derived from the use of melted wax which was practised in Egypt in the 3rd century. Oil-painting derives from the progress which was being made in **chemistry**, which enabled **purified linseed oil** to be obtained, and which then produced **volatile solvents**. It was three or four decades later, around 1450, when the use of oil as a medium became generalized, following the technique being introduced in Venetian painting by Antonello da Messina.

Strangely enough, it was only in the 19th century that the oil + pigment combination became an industrial concern.

Organ

Ctesibius, 3rd century BC

There is sufficient evidence to be able to say that Ctesibius, a Greek mechanic from the Alexandrian School who lived in the first half of the 3rd century BC, built a **hydraulic organ**. Although it has been incompletely described by Philo of Byzantium, Hero and Vitruvius, as well as by various Arab manuscripts, it has been possible to reconstruct the instrument schematically in the hands of specialists such as Bertrand Gille, based both on ancient representations and on evidence discovered in Naples and in Hungary. As well as **sound pipes with reeds**, the instrument included a **wind-chest**, which stored air, and **mechanical bellows**, which were powered by a **hydraulic suction and force pump** (see p. 108). In addition, it obviously comprised a keyboard which activated **valves** to open or close whichever pipes the air was directed towards. This was undeniably the first mechanical musical instrument, and it could play for as long as the bellows kept going. In the Middle Ages, the organ-makers began to diversify their manufacturing methods.

Some ancient documents indicate that Ctesibius's wife, whose name was Thaïs, learned how to play the instrument which her husband had made. She would therefore have been the first organist in history.

Parasol

Anon, China, 4th century

The first parasols appeared in the 4th century in China under the Wei dynasty. They were made of oiled paper stuck on to flexible ribs and were used for protection from the rain as much as from the sun. They were therefore designed on the model of the **umbrellas** which had preceded them by about four centuries.

Pen

Anon, Egypt, c.5000 BC

The ancestor of the fountain-pen seems to be the shaped reed which was filled with ink and which existed in Egypt at least five centuries ago; it sufficed to press lightly to make the ink run down to the point. It is not known how the Egyptians controlled the fluidity of the ink. A shaped **bamboo** which was filled with ink in a similar manner was used in China. It is not known whether the 'silver reed' used by Aristotle in the 4th century BC had a **reservoir** in it or not, but it is known that during this philosopher's time there were metal fountain-pens, as attested by the bronze pen found in Pompeii which is now exhibited in Naples Museum. The fountain-pen was found in 17th-century Paris under the name of the 'endless pen'. It was made of silver and cost 10 to 12 F at the time. Renamed the '**inkpot-pen**' in the 19th century, the instrument was patented twice in succession, once in 1864 in France, by Mallat, and again in 1884 in the United States by Lewis Waterman.

An 'endless pen' dating from the time of Louis XVI and made out of ivory and gold was sold in Paris at the Hotel Drouot in 1954 for 19 500 F.

Piano

Hero of Alexandria, 1st century BC; Vitruvius, 1st century; anon, Italy (?), 10th century; Gui d'Arezzo (?), 11th century; anon, Germany, 1440; Spinetti, 1503; anon, Europe, 1700; Cristofori, 1720; Schroeter, 1763; Zumpe, 1765; Stein, 1780; Erard, 1821; Boehm, 1831; Steinweg, 1853 etc

The piano represents the perfect example of a continuous invention, which adapts itself ceaselessly to the evolution of culture. Far from having appeared suddenly, it proceeds from a long line of inventions stretching over about twenty centuries, the rhythm of which accelerated as the instrument and its use became more widespread and its technique diversified.

Its basic principle, the **percussion**, rather than the plucking, of a string by a key which raises a wooden peg starting at the bottom, is surprisingly old, for it is described in Hero of Alexandria's *Pneumatics*. It is not known whether Vitruvius, the Roman architect and inventor who lived in the 1st century BC, took the idea of the **keyboard** from Hero or if in fact he re-invented it, but Hero is known to have been the first to design the **key**. Whether these percussion and keyboard instruments existed in antiquity, we do not know, for none have survived.

The piano's oldest ancestor, which actually was built and can be dated, was the **organum**; this is thought to have been made, if not invented, in 10th-century Italy. According to some later representations which have survived, it was an oblong box of about thirty centimetres in length comprising **fourths** and **octaves** and it was destined to provide an accompaniment to the human voice below the melody. The more bulky **organistrum** appeared at approximately the same time; this was the **hurdy-gurdy** from which the barrel-organ was derived, and which produced two kinds of sounds: in addition to the percussion, three strings rubbed by a crank-activated wheel produced a **continuous bass**.

In the 11th century there was a revolution in music, the introduction of a numerical system of transcribing notes, which was inspired by Pythagoras and invented by Gui d'Arezzo. At this time an instrument which was essentially pedagogical in use appeared, the **monochord**, which, as its name indicates, consisted of a single string which was stretched over a horizontal case and supported by a movable bridge. The purpose of the monochord was to show the proportions of the intervals between notes to music students and to teach them tonal gradation. It was perhaps Gui d'Arezzo who invented it, but this is not certain. The fact remains that half a century later a version with four strings and a keyboard appeared, called the **quadriplex monochord**, the scales of which show the four authentic modes and the four plagal modes of plainsong, high-pitched fifths and low-pitched fourths.

A drawing dating from 1440 which is kept at Weimar depicts an instrument which heralded the harpsichord, the **clavichord**, which seems to have had sixteen long keys and eight short keys, as well as twelve strings. The bridges supporting the strings were obviously fixed. The instrument seems to have evolved from the end of the 13th century, or in any case the beginning of the 14th, as is attested also by a mention of it in the Minnesingers' rules, dating from 1404.

Some writers tend to think that the clavichord antedated another instrument which constituted one stage towards the piano, the **clavicimbalum**, a version of which is thought to feature in *The Mystical Lamb* by the painter Van Eyck; this was a kind of

> Mozart played a major role in the development of the modern piano, for he lavished his approval on Stein's invention. His own grand piano, a Huhn, had been made in Berlin exactly according to Stein's specifications.

positive organ, and the instrument resembled a keyboard harp. On one panel of the famous triptych depicting St Cecilia, the angel has his right hand on the **la** of a minutely drawn keyboard, while his left hand rests on the **do** and the **fa**. However, nothing of the mechanism of the instrument is known.

At the beginning of the 16th century, the clavichord attained a certain degree of complexity; it was a large box with a lid, nearly a metre in length, which included nearly eight octaves. As well as the clavicimbalum, there were two other versions of it, the **virginal** and the **clavicitherium**. It seems that in these four instruments the engineers did not free themselves from the model of the monochord, for the length of the strings was the same for all of them, with their frequencies being altered only by the spacing of the bridges.

It was in 1530 at the latest that the Venetian Giovanni Spinetti invented an instrument where the strings were for the first time, of decreasing lengths; this was the **spinet** and it was trapezoidal in shape. Since they were built in different sizes, and so with different numbers of octaves, they were sometimes superimposed in concert in order to achieve outstanding effects of *piano* and *forte* in juxtaposition. The largest spinet is the **harpsichord**.

In 1720 the Paduan Bartolomeo Cristofori was the author of a truly brilliant invention. First of all he put the spinet flat on three feet and fitted it with a single 29-key keyboard, which could play both **piano** and **forte**. Then he made a mechanism of **escapement percussion**, which meant that the key no longer pinched the string as it did in the harpsichord, but hit it with a leather-covered hammer, by means of a pivoting lever. The sound obtained was totally different, being without the harmonics of the harpsichord. It followed that the playing could be more expressive, since the strength of the note depended on that of the impact on the string. This was the **pianoforte**.

In the beginning the instrument must have been far from perfect, for when Voltaire compared it to the harpsichord he consigned it to the boiler-makers! Listening to a perfectly restored period *pianoforte* can be irritating for the 20th-century listener because of the clicking of the falling keys. Consequently many inventors endeavoured to improve Cristofori's invention: among the most interesting are the work of the Saxon Schroeter, which reduced the track of the lever, but also absorbed the vibrations with its system of shock absorbers, and that of a German living in London, Johann Zumpe, who modified the shock absorption in order to give the note its harmonics.

The pianoforte can be considered to have passed away and been succeeded by the piano when the German Johann Andreas Stein invented a mechanism which managed to free the key, the path of which had until then been limited by the pivoting lever. His new mechanism, which was called 'Viennese' gave a very much more elastic contact on the string; it appeared in 1780 together with the installation of a **mute** called **'celestial'**.

In 1776 the Alsatian Sébastien Erard built the first rectangular piano, and in 1808 he had the idea of giving the strings more elasticity with a new system of bridges. In 1821 Erard equipped the percussion mechanisms with springs which brought the hammer back into position immediately after the note had been played, and thus gave the player much more flexibility than before.

Ten years later the German flautist Theobald Boehm, taking advantage of the fact that the sounding boards were now made of metal, thought of increasing the tension of the strings to achieve more tonal clarity.

In 1853 the German Heinrich Engelhard

Care should be taken with regard to the former vague uses of the names of the instruments described above, for in the 16th century some writers used the following terms in an indiscriminate way: **'virginal'**, **'spinet'**, **'harpsichord'**, **'clavicimbalum'**, **'psalter'**, **'octavine'**, etc. This was firstly because their definitions and then their conceptions had not been codified. Furthermore, some manufacturers modified their instruments in what was often an unprecedented way. Consequently when Johann Christian Bach arrived in London in 1759, some manufacturers of second-rate harpsichords tried to transform them into pianofortes, without much success.

Steinweg, who lived in New York and was the founder of the famous Steinway firm, introduced his grand pianos with highly tensioned strings and large sounding boards, the harmonic fullness of which was to attract countless musicians.

There are estimated to be more than two thousand inventions in the world from which the making of the piano has benefited. Each firm introduces versions and improvements in various components of its instruments, in such a way as to confer upon them a distinct sonority and specific qualities.

Playing cards

Anon, China, 9th century

Playing cards appeared in Europe during the last quarter of the 14th century and were invented in China during the 9th century. The first explanation of a card-game was written by a woman.

Public lighting

Anon, Rome, 1st century; Paris Parliament, 1525; La Reynie, 1667; Paris Parliament, 1704; Bourgeois de Châteaublanc, 1766; Lebon, 1786–1799; Murdoch, 1792

It was in the Roman Empire and under the Augustine influence in the 1st century that public or municipal lighting seems to have become a reality for the first time. Until then only the wealthier people could see outside at night as they could have torch-bearers going along in front of them. However, torches stuck into iron shafts fixed to the buildings to light the main avenues and streets of the towns of the Empire lent a somewhat extravagant hue to this kind of equipment. The Roman system, which ensured illumination throughout the night, declined with the Empire and then disappeared altogether. During the Middle Ages in Europe and apparently in the rest of the world as well, public roads after nightfall were abandoned to the darkness and the cut-throats. In 1525 a decree from the Paris Parliament, apparently the first of its kind in the West, ordered the wealthy classes to put a candle at their windows to throw a little light on to the street. This arrangement seems to have had little effect, however, for in 1559 a new decree ordered the installation of lanterns at the street corners, at the cost of the same bourgeoisie. Their effect seems to have been disappointing, and the carried torches reappeared. In 1667 the Parisian police lieutenant La Reynie, whose office had just been created, imposed lighting using tallow candles, still at the cost of the wealthy. Until 1704 the five thousand candles which illuminated the main roads of Paris — this is not counting the provincial towns — were paid for by these middle classes; that year the Paris Parliament decided to bear the cost itself.

> The first town in the world to be lit using municipal gas was in fact Freiberg, which is now part of Germany. This was owing to the work of Wilhelm August Lampadius, who was also the inventor of the first gas omnibus. In 1811 several streets of this Saxon town had free gas lighting.

In 1766 Bourgeois de Châteaublanc saw his invention of **street lamps** become widespread as a result of help from the police. These were a novelty, a unique model of lanterns on stands with an **oil lamp with a flat wick** and a reflector. In 1782 or 1783 the Swiss Aimé Argand invented the **double air-flow oil lamp** which had no annoying smoke, and which was quickly manufactured in England by Matthew Boulton.

The first modern municipal lighting network was invented by Philippe Lebon, who discovered **coal gas**. Having obtained an unknown, combustible gas in 1786 (which was in fact **hydrogen**), by the distillation of wood or charcoal, Lebon first of all invented the '**thermolamp**'. In 1801 he was given the task of illuminating the celebrations for the imperial coronation. Perhaps he did not sufficiently protect the secret of how he obtained coal gas, for in 1792 the Scot William Murdoch discovered it and planned to exploit it, but he came up against the competition of the German Friedrich Albert Winzer, or Winsor, who, having been unable to procure Lebon's patent, had copied it in England. In the year of Lebon's mysterious assassination, Winsor submitted a patent for the manufacture of gas from coal. Then Winsor founded the Chartered Gas Light and Coke Company and in 1813 the first area of London was lit with municipal gas, that of Westminster. It was not until 1829 that Paris began to benefit from what was, after all, Lebon's invention.

Roller-skates

Anon, Holland, 18th century; Merlin, c.1760; Plimpton, 1863; Richardson, 1884

The **ice-skate** existed in Scandinavian countries in the first centuries AD at least, and it may be even older. The oldest evidence of its use dates from the 8th century, and in the 15th century there were skates which were set either on a **metal blade** or on a **bone blade**. The skaters' patron was Saint Lidwin, who had a skating accident himself in 1396. It was only in the 18th century that an anonymous inventor, who is thought to have been Dutch, created the '**ground skate**'. This consisted of a rigid sole with four wheels which was attached to one's shoe.

The first known manufacturer of roller-skates was the Belgian Joseph Merlin from Huys who had a shop in London around 1760. During the ensuing years the *patin à terre* appeared in France, as did the *Erdslittschuh* in Germany. Travelling round corners was difficult with these skates and several times attempts were made to remedy this by replacing the four wheels by **two wheels**, following the model of the ice-skate, and despite several failures in these efforts, skating became increasingly popular.

Skating rinks were built in several large towns in Europe and America and skating entered the history of art in the Paris Opera: this took place in the ballet of Giacomo Meyerbeer's opera *The Prophet*, where the ballerinas appeared on roller-skates, enchanting the audience with their swift movements. The famous ballet-master Paul Taglioni, brother of the ballerina Marie Taglioni, even created a ballet which was appropriately entitled *The Skaters* in which ice-skaters were actually represented by people wheeling round on roller-skates.

In 1863 skating was given a new boost when the American James L. Plimpton invented **shock absorber skates**. In 1884 the use of **ball-bearings** in the wheels, invented by another American, Levant Richardson, lent a fluidity of movement to the skates which helped to increase their success.

Sewers

Anon, Assyria, 8th century BC; Rome, 6th century BC

Originally the invention of sewers was limited to the installation of **gutters** which were meant to carry along both rainwater and waste-water. These gutters were dug in the centre of the streets in the towns of the second Assyrian Empire, apparently in about the 8th century BC, although the actual date is not certain. The Greeks dug sewers too, but it was the Romans who, with the methodical spirit which they applied to all their public works, were the first to install **underground mains sewers**, such as the famous *Cloaca Maxima* which was built in the 6th century BC under Tarquinius the Elder. The *Cloaca Maxima*, moreover, was meant to drain the ground under the Forum. It would seem that in France drains or gutters were dug directly into the street soil, until Philippe Auguste's decision in about 1200 to pave the streets. The first **covered stone sewer** was built in 1370.

Soap

Anon, 2500 BC

The origin of soap is obscure. A Sumerian text dating back to 2500 BC described the manufacture of a **detergent** product, based effectively on **fat** and **alkali**. Soap however did not exist in Greece, where saponin juice was used, nor in Rome during the first centuries, where an **abrasive paste** based on ash and fat was used. The manufacturing technique based on animal or vegetable fats and alkali was introduced into Europe by the Teutons in the early Middle Ages. Nevertheless it is worth noting that the potassium carbonate in plant ash would have acted like an alkali in the Roman paste.

The manufacturing techniques of soap were improved considerably owing to the completion of the process for obtaining caustic soda by **treating sea-salt with sulphuric acid**, achieved by Nicolas Leblanc in 1791. They also benefited from the discovery of **olein** by Michel Eugène Chevreul in 1823 and then the use of the Solvay process for making **ammoniac soda**.

Spectacles

Anon, c.300 BC; Armati or Spina, c.1280; Franklin, 1784

The origin of **corrective spectacles** which is generally given by historiographers points to a man named Alessandro della Spina, who exploited the invention of a friend, Salvino degli Armati around the year 1280 in Florence. The first glasses to be made in this way had **convex lenses** and so would have served to correct **hypermetropia** or **longsightedness**. In reality, this version of the origin of spectacles is doubtful. Even though it is correct that no traces of corrective spectacles dating from before the 13th century have been found, it is also the case that old glass lenses have been found dating from 2000 BC, lenses were used either for making fire or as magnifying glasses. In Aristophanes's comedy *The Clouds*, which dates from the 5th century BC,

Spectacle manufacturer's *workshop in the Middle Ages. Engraving by Philippe Galle (1537–1612)*

there is mention of a crystal being used to light the fire, and it is unlikely that the generations of people who used this instrument, which probably consisted of shaped and polished quartz, were unaware of its optical properties. Logically speaking, they would have then been able to set lenses either individually or in pairs, in order to correct faults in vision, beginning in fact with long-sightedness.

The fact remains that corrective glasses appeared quite mysteriously, both in Europe (in Italy first of all) and in China, for which no precedent can be established, and that they were first used for the correction of longsightedness. In 1352 the Italian painter Tommaso da Modena painted the portrait of Hugues of Provence wearing spectacles, which would suggest that these accessories had become relatively common by the 14th century.

However it was the beginning of the 16th century before **myopia** or **short-sightedness** was corrected using concave lenses, such as those worn by the pope Leo X in his portrait by Raphael in 1517.

The explanation for the relative rarity of spectacles in the past centuries is undoubtedly to do with the fact that the 'glasses' were shaped out of precious stone, either **quartz** or **beryl**, and that there was also the difficulty of polishing them. Moreover, the correction of vision defects must have been approximative, given the small amount of information known about **optics** and especially **optometry** at the time. It was only when the glasses actually began to be made of **glass**, generally from Venice and Nuremberg in the 16th century, that spectacles began to be widely available.

The largest step taken previous to the advent of optics and then optometry was the making of **bifocal glasses**, invented in 1784 by the American Benjamin Franklin. At the time these consisted of distinct segments of lenses which were held together by the frame.

> Some second-hand information, from late historiographers, reports that at the time of the Arab conquest the Alexandrian lighthouse was equipped with a telescope and Nero wore a monocle made out of an emerald, or a green beryl, to correct a defect in his vision.

Trade unions

Anon, Europe, 14th century

It was within the guilds which arose in the 14th century in Europe that workers first developed organizations against their employers, the master craftsmen, in order to defend their rights. The phenomenon was particularly marked in Great Britain, where it apparently created a tradition, for in 1720 the first professional unions formed and the first strikes took place, as attested by the complaints of the master-tailors of London to the Parliament. The first general trade union was founded in 1833 under the influence of the Welshman Robert Owen; this was the Great Consolidated Trade Union, the first to obtain legal status.

Wig

Anon, Egypt, 3000 BC; anon, England, 16th century

The first known wigs seem to have appeared in Egypt in 3000 BC; they consisted mainly of **ornamental coiffure**, made using tapered and plaited reeds, rather than **false hair-pieces**. Assyrians, Phoenicians and Romans all used them, but they disappeared after the fall of the Roman Empire and reappeared, anonymously, in 16th-century England. The fashion for them seems to have been launched by Queen Elizabeth I who was congenitally bald. The portraits of the monarch show that wigs at that time consisted of postiches, designed to disguise the natural absence of hair, and were made of real hair or hairs of animal or vegetable origin. The fashion slowly took over France as well. It was imposed by Louis XIII, who was the first king of France to wear a wig, and it disappeared at the Revolution; when wigs reappeared in the 19th century, they were used exclusively as false hair-pieces.

Energy and Mechanics

After a study of the great inventions which flourished in the twin domains of energy and mechanics, it is tempting to conclude that mechanics was mastered at an early stage — but this did not lead to a development of technology. The Greeks and the Chinese were prodigious mechanics; it is astonishing to discover that there were fire-engines in the ancient world (yoked to animals of course) which comprised both cylinder and piston. Reaming too was carried out quite successfully. Automatons, regulators, chain transmission, the universal joint, the translation of circular movement into rectilinear movement and demultiplication: all these are known to be very old, their inventors often unknown. It was not until the Renaissance that inventions, like works of art, began to be signed on a regular basis.

This mystery, the long delay which separates mechanics from technology, could be explained by the time taken before the eventual mastery of steam-power. However, Hero of Alexandria had shown in the 1st century that steam could provide energy.

It took the human race hundreds of thousands of years to progress from nomadism, hunting and gathering to settlement, agriculture and stock-rearing, but they needed a little less than two millennia to pass from simple ways of using energy, notably in the windmill and water-mill and propulsion by sail, to the mastery of energy. Then everything happened at once; hardly a few decades separated the mastery of steam from that of electricity. Humankind eventually reached the age of reason during the Industrial Revolution and, often without knowing it, re-invented the ancient wonders of the human mechanical genius.

Air and water cooling

Stirling, 1816; Forest, 1881

The idea of using the **circulation of water** to cool an engine seems obvious, but it was apparently not thought of until the Scot Robert Stirling applied it to his engine (see p. 107). Its first large-scale use was in the principle of the **radiator**, and in this application consisted of a coil filled with water and placed near the engine; this goes back to the beginnings of the **car industry**. **Air cooling** was invented in 1881 by the Frenchman Fernand Forest who came up with a remarkable solution: small currents of air circulated in cylindrical helicoidal grooves, described as fins, which were formed in the cylinder when it was cast.

Atmospheric engine

Papin, 1687; Newcomen, 1698; Smeaton, c.1775

An atmospheric engine is a machine where **atmospheric pressure** is exerted complementarily to the pressure of **steam**. It has a short but interesting history, extending over barely one century; its beginnings are mixed up with those of the steam engine. It was Denis Papin in 1687 who by inventing the first steam engine of modern times also invented the first atmospheric engine. It consisted of a cylinder in which the expanding steam pushed a piston upwards; when the steam condensed the piston was driven down by atmospheric pressure.

The first to make an atmospheric engine was the Englishman Thomas Newcomen, who is thought to have been told about Papin's invention by Robert Hooke. Built in 1698 with the help of John Calley, who lived in Dartmouth as did Newcomen, this machine was met with reservations, firstly because it was not Newcomen's invention, and secondly because it was not 'original', in the sense that it included no really new mechanical concept. The fact remains that it is significant in history because it was the first one of its kind to have existed. It consisted of a cylinder placed above a boiler and into which a stream of water was injected; this vaporized under the effect of the heat and, as we have seen above, this steam then lifted a piston. After **condensation** the piston came down again. This piston was linked to one of the two ends of a **beam**, and as the piston fell the other end of the beam rose. The machine was to be used to pump water out of a mine. Its motive force was therefore dependent on atmospheric pressure.

The first model made by Newcomen had some technical difficulties such as problems of welding and adjustment. The first satisfactory atmospheric engine was put into use in 1710; its power was about 5.5 hp. After being improved by Henry Beaton, who changed the cooling system in the cylinder to work using the circulation of cold water, and also benefiting after 1730 from improved **cast-iron techniques**, which enabled robust, watertight cylinders to be made cheaply, the atmospheric engine was very successful, not only in England but also in the rest of Europe. It had the great advantages of drawing the attention of scientists and engineers to **heat engines** and preparing the way for the advent of the actual **steam engine**.

Around 1765 the Englishman John Smeaton became interested in it, and around 1775 he increased the output considerably by replacing the beam with an **oscillating wheel**, by increasing the **stroke of the piston** and by calculating both the optimal speed and the useful rate of cooling.

During the winter of 1763–1764 the Scot James Watt, a professor at Glasgow University, was called to inspect an atmospheric engine which was not working. With his methodical mind, Watt noted that the volume of the steam inside the cylinder at atmospheric pressure was 1800 times more than that of the same vapour in its condensed state, which implied a disproportionate expenditure of energy, thus preventing the continuous working of the machine. The loss was even greater because this steam was being condensed by the cooling and therefore losing its calorific energy. He thought of salvaging the steam by adding a **condenser** to the engine into which it would be sucked, because a **vacuum** would have been created inside. In this modified version of Newcomen's engine, there was firstly economy of heat, since the steam was no longer cooled but cleared. Secondly the piston was no longer driven down by atmospheric pressure but by induced pressure; the re-injection of the steam made the cylinder rise again. The atmospheric engine was replaced by the steam engine (single-effect type), and Watt's brilliance had given a premonition of the **ideal cycle** which Carnot was to define almost exactly half a century later (see p. 85).

Automatons

Archytas of Tarente, c.350 BC; Philo of Athens, c.250 BC; Hero of Alexandria, 1st century BC; Vaucanson, 1738

The first automaton, understood in the modern sense of the word, appeared very early in history with the artificial flying bird of the Greek Archytas of Tarente. This dates from around the middle of the 4th century BC. It very probably consisted of a bird made out of light wood which was fixed by a wire to the end of a light wooden shaft; with its jointed wings, the device would have given the impression of flight when a steam jet was directed towards it. This is the first known attempt at the mechanical reconstruction of a biological movement. One century later, Philo of Athens was to make the first proper automatons, such as a statue of a slave pouring wine. His instruments undoubtedly worked using a system of **hydraulic circuits** and **counterweights** activating **levers** and they were designed following a similar principle: pressure of water or steam lifted a float which was attached to a lever. A range of variations depending on the ingenuity of the designer allowed the gestures to be broken down at will. An example of this is found in a description of the **clepsydra** by Philo of Athens. This consisted of a set of three copper vessels which fitted inside each other and were connected by pipes which exchanged the water continuously by the following system: when the water reached a certain level in one, a valve stopped the flow so that the water flowed into another vessel. Although this was meant as a decorative clepsydra, the mechanism is still the first known example of **feedback**.

Another example of feedback, which is at least as surprising for the time, is given by a Chinese invention which will be referred to under the rather complicated name of **wise man's south-pointing chariot**. It consisted of a chariot, 3.3 m long and 2.85 m wide, which had **differential gearing** inside it and a statue of a sage on top; the outstretched arm of the sage always pointed towards the south. Whatever the direction taken by the chariot and the number of successive turns it made, the outstretched arm invariably remained pointing in the same direction. This prodigious vehicle would have been built to the order of the leader or duke of Chou (Zhou) at the beginning of the 1st century BC, in order to bring envoys back safely and quickly.

This vehicle gives rise to two fundamental questions. The first concerns the mechanism which made it work (this did not involve a magnet); the second is its date of manufacture. The mechanism has been rebuilt, hypothetically, for we have no technical description of the time but only a rather vague written description. Its modern mechanic was G. Lancaster, and the vehicle is now exhibited in the London Science Museum. The mechanism consisted of various sizes of cogwheels which made up a **gear-train** which, as in a contemporary motor car, transmitted to a **rotary shaft** the sum or the difference of two other movements; in a motor car the **differential** enables an outside wheel to have a faster speed going round a corner than an inside wheel. In the Chinese chariot, the gear-train would have been sufficiently precise to transmit the difference in the relative movements of the wheels at each corner to the shaft supporting the statue and consequently to enable it to correct its relative angle. The invention is plausible; the Greeks also invented the train of differential gears (see p. 89) in the 1st century BC.

Despite the fact that its authenticity is defended by the authority of Joseph Needham, the specialist on Chinese technology, there is nevertheless a problem with the invention. This is not because there is any doubt about the ingenuity of the Chinese inventors — at least one of them made the **universal joint** (see p. 112), a feat no less astonishing and this time absolutely certain — but for the following reason: it presupposes a totally astounding, if not improbable, level of mathematical and technological knowledge for the beginning of the 1st millennium BC. Needham himself doubts this date and only holds that of the 3rd century as definite, when a different reference attributes the invention of the chariot to a remarkable engineer, Ma Chün. In order to function correctly the differential of the chariot could not be subjected to errors of more than 1%, which hardly seems compatible with the state of Chinese metallurgy 3000 years ago. On the other hand, it is hard to see why in the 3rd century, when the **compass** had already existed for two centuries (see p. 154), an engineer, however skilled, should tackle an undertaking as formidable as that of a differential, which if it wore out was likely to lose the duke of Chou's (Zhou) envoys.

This chariot has given rise to numerous commentaries, which do not all show the good sense desirable in a history of technology. Its supporters evoke with no more exactitude the description which has come to us of another mobile craft, a boat, which was built under the Chinese dynasty Chin (Jin; 265–420); it sailed for the pleasure of the court on a lake near the Imperial Palace and it too carried a figure with its arm outstretched in one direction. Needham admits that this figure could have taken its direction from a magnet, but assumes somewhat rashly that if the boat was powered by **paddle-wheels** (see p. 202), it could equally well have included a differential. It is difficult, however, to imagine a differential of this level of precision between two paddle-wheels.

Finally there remains the hypothesis that the Chinese discovered the differential through the Greeks, and that they adapted it to a mechanism which was of mysterious appearance and philosophical inspiration, as is the case with several of their inventions. Their technological genius has proved itself sufficiently for the question of the *wise man's south-pointing chariot* to remain open until further information is available. Whether he owed his positioning to a magnet or a differential, this sage was undeniably an enigmatic automaton.

The fact remains that the production of automatons was increasingly popular from the 1st to the 19th centuries, particularly with monarchs such as Emperor Theophilus of Byzantium (who reigned from 829 to 842), whose throne was decorated with mechanical lions which shook their heads and roared while mechanical birds chirped ... One of the most famous of these automatons is *Charles V's ship*, which is in the Cluny Museum in Paris. It is a clock in the shape of a boat with working **organs** which sound the hour whilst musicians play their instruments, sailors bustle about and courtiers pay their respects to the monarch, who nods his head.

Hero of Alexandria gives an account of a notched component in the form of a cylinder. Each notch generated a different movement of the part, in this case a **cogwheel**, to which it was connected. This was in fact the forerunner of the modern **camshaft**, which was used in the construction of numerous automatons, especially those dating from the late Middle Ages and later.

It was almost certainly the camshaft which inspired Gerbert (see *Clocks*) to come up with the principle of **escapement**, for he was the first to apply it to clockmaking. The camshaft developed considerably during the 18th century, with the work of Jacques Vaucanson who made it compete with the **punched card**: this device became a cylinder bristling with spikes which were arranged at predetermined locations and had particular heights, in such a way as to engage gearwheels and levers at certain times to obtain precise movements. It was in this way that in 1738 a masterpiece was made, the *Player of the transverse flute*, which could play twelve different tunes. Having been invented in 1725 by Bouchon (see *Weaving looms*), the punched card itself was taken up again and improved in 1745 by the same Vaucanson, for use in **weaving looms**. From that moment onwards the industrial branch of automatons had begun. In 1789 the Scot James Watt, totally independently, was to patent one of the precious inventions of Vaucanson and his predecessors, the **centrifugal regulator**. This consisted of a weight fixed to an articulated shaft which turned by coupling with the axle of a motor (a **steam engine**). The acceleration of the engine caused the faster rotation of the weight, which was then diverted from its original plane of rotation by the **centrifugal force**. Thus the pressure of the articulated shaft on the steam inlet valve was controlled; this therefore tended to slow down the speed of the motor and maintained it within certain limits. This was the first verified example of **negative feedback**. Automatons then gave rise to automation, an example of which is **computing**.

Ball-bearings

Anon, Mesopotamia, Egypt, 3000 BC; anon, Rome, 1st century BC; Leonardo da Vinci, c.1500; Michaux, 1862

The principle behind the ball-bearing can be said to date back to the **rows of logs** used around 3000 BC in Mesopotamia and in Egypt for the transportation of heavy objects, such as blocks of stone or boats. It is a mechanical principle which uses balls to reduce the friction between two moving objects at the point of contact. Evidence of it is found in Greece in the 5th and 4th century. Roman remains discovered in 1928 during the draining of Lake Nemi in Italy showed that Roman engineers invented a sort of primitive ball-bearing, consisting of a cylindrical case containing bronze balls. This case might possibly have served to reduce the friction between metallic objects and wooden objects, after the fashion of the metal disc designed by Hero of Alexandria for reducing the friction between the revolving parts of **catapults**. The first systematic study of friction seems to have been done by Leonardo da Vinci, whose *Notebooks* contain numerous designs of ball-bearings, and whom we can consider to be the progenitor of the modern concepts in this field.

The arrival of modern means of transport and improvements in metallurgy in the 19th century stimulated interest in ball-bearings. Very hard steels meant that strong balls could be made. The **bicycle** was the first vehicle to benefit from the installation of ball-bearings, in 1879. Ball-bearings seem to have been first patented in 1862 by the Frenchman Pierre Michaux.

Carnot cycle

Carnot, 1824

Sadi Carnot's invention was a major event in the history of both science and technology because it founded a new science, **thermodynamics**, which was to have a profound effect on physics and mechanics. It was in 1824 at the age of 28 that Carnot, the son of one of Bonaparte's war ministers, published his *Réflexions sur la puissance motrice du feu et les machines propre à développer cette puissance.* (Report on the driving force of heat and the correct machines to develop this power). In it he proposed to outline a theory of the **steam engine**. The wonder of his invention is that it was based on a false concept: Carnot, like his contemporaries, thought of heat as a fluid, the **caloric**. This notion was an avatar of the theories on **phlogistics**, which Lavoisier had in fact disproved in 1777 with the discovery of oxygen. Physics then was only in its infancy and the concept of **particle energy** was still three-quarters of a century away (it was actually not to appear until the beginning of the 20th century when the theories of the German Ludwig Boltzmann had been assimilated). Carnot therefore thought of the steam engine as a type of mill where the flow of the caloric made the parts move. Carnot deduced that there was conservation of the caloric in the same way as there was conservation of water in a mill, with the heat passing from an upper to a lower level, but that the cycle of the caloric had to be interrupted for there to be work. It was on these bases that he described the theoretical cycle perfectly.

In order to understand how the ideal machine of the Carnot cycle works, certain points must be clarified. When a piston compresses a gas in a cylinder, the temperature of this gas increases; the compression is therefore said to be **adiabatic**. If the cylinder contains a certain mass which absorbs its surplus heat its temperature does not vary and the compression of the gas is called **isothermic**. If the gas contained in the cylinder is isolated and so no change, either positive or negative, takes place with the external environment, the compression is adiabatic. In adiabatic compression, the pressure increases more than in isothermic compression; this is due to the initial pressure bringing about additional pressure because of the heat.

Carnot's idea was that in a cylinder in contact with a source of heat, such as in the **Stirling engine** (see p. 107), which would later give rise to the Carnot cycle, four

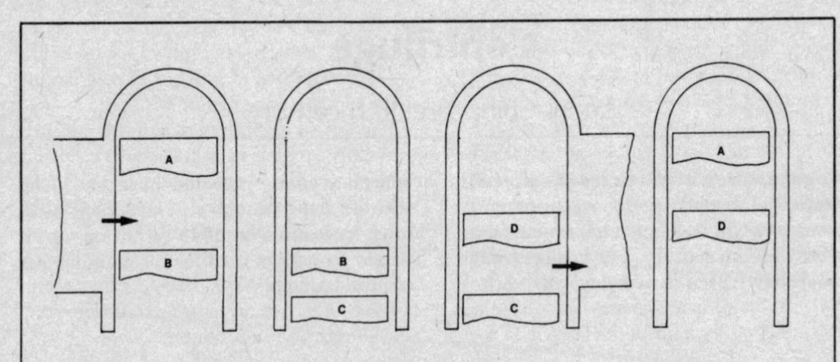

***The Carnot cycle** comprises four stages. From A to B, the expansion is isothermic, from B to C, it is adiabatic; these two stages are motive. From C to D, the compression is isothermic, from D to A, it is adiabatic; these last two stages consume energy. Each stage corresponds to an almost static transformation*

stages must be distinguished. The first is where piston A is at the end of its stroke due to the expansion of the gas; this is an isothermic expansion since the external source of heat makes up for the inherent loss of heat in the decompression. In the second stage, from B to C, the end of the expansion is adiabatic, for there is cooling of the gas as a result of the decompression. In the third stage, from C to D, the compression is isothermic, and in the fourth stage, from D to A, it ends on an adiabatic phase. It is the compression stages which use up energy, and the two expansion stages which produce it. Carnot postulated that the energy generated during the motive phase does not completely restore the initial conditions of pressure, so that the second compression consumes less energy than has been generated by the expansion. Carnot therefore proposed to cool the gas at a given point of the expansion, the point B where the gas had reached its maximal point of expansion and generated all of its power, beyond which the expansion had to end with an adiabatic phase. In this way, the compression from C to D, which would be effected on a cold gas, would require less effort. This would therefore be an isothermic compression. Here, the contact with the cold source would be cut and the compression would end on a heat-producing phase in order to produce the energy for the following cycle; this final phase would therefore be adiabatic. Carnot actually divided compression and expansion into two parts, the one isothermic and the other adiabatic.

The lesson behind this demonstration was that in order to obtain energy provided by a hot source, part of this energy must be ceded to a cold source. The use of energy does not happen unless there is a disequilibrium, without which the energy cycle is entirely isothermic and closed (this is known as the **Atkins cycle**).

This original machine constituted a demonstration of that which was to become the second law of thermodynamics (the first being that of the conservation of energy, that is, 'Energy is neither created nor destroyed').

It is worth noting that it was the Frenchman Emile Clapeyron who gave a finer analysis of the Carnot cycle in his *Report on the motive power of heat* in 1834 which drew the attention of the science world to Carnot's brilliant invention. The influence of the Carnot cycle on the exploitation of the steam engine and on all machines (those which imply a transfer of energy of course) has lasted to the present day. It has had a particular effect on the Stirling engine (see p. 107) which constitutes one of its purest applications and the construction of which did not begin to be mastered until the second half of the 20th century.

Centrifuge

Anon, China, pre-10th century

The **separation of particles of different densities** by rapid rotation was apparently first used in China, at an indeterminate date before the 10th century. The first centrifuge would have been a wooden tub with a conical section. It would have had a lid with an axle through it, two spoons at its lower end and a crank to turn it on top. It would have been used for the manufacture of lubricating oil.

Centrifugal pump

Leonardo da Vinci, 1508

Among the numerous technological projects described in Leonardo da Vinci's *Notebooks* there is one dated 1508 which seems to constitute the first known prototype of the **centrifugal action pump**. It consists of a tub of conical section at the centre of which there is a fixed axle. This has a propeller at its lower end and at the top end there is a bit-brace linked to an energy source such as a water turbine. The movement of the propeller would cause the formation of a swirling depression in the shape of an inverted cone, which would make the water rise up towards the edge of the tub. The water could then be emptied by a pipe situated — in the drawing — a short distance from the edge of the tub. A model was made of the appliance for some German engineers. It was to be used for **draining marshes**.

Chain transmission

Philo of Byzantium, c.200 BC; Chang Hsu-hsün (Zhang Xuxun), 976; Vaucanson, 1770

The oldest description of a transmission of movement between two cogwheels by an endless, linked chain is that by the Greek Philo of Byzantium and it dates from 200 BC; it was designed for loading catapults. In 976, Chang Hsu-hsün (Zhang Xuxun) made a similar chain for his great astronomical clock (see p. 153), calling it a celestial ladder. The principle seemed to have been lost from the time of Philo of Byzantium, but this time it survived, for the inventor Su Sung, also Chinese, left behind a very precise drawing of this transmission system which is dated 1094. Was the invention lost again? Everything would seem to suggest so, for it reappeared only in 1770, owing to the Frenchman Jacques de Vaucanson who used it in a new **power loom** (see p. 138).

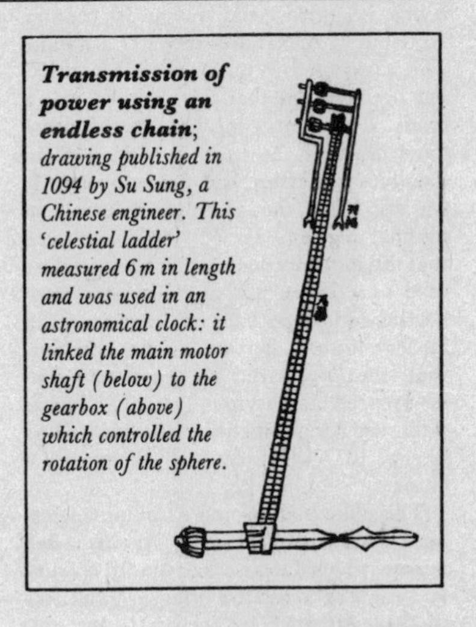

Transmission of power using an endless chain; *drawing published in 1094 by Su Sung, a Chinese engineer. This 'celestial ladder' measured 6 m in length and was used in an astronomical clock: it linked the main motor shaft (below) to the gearbox (above) which controlled the rotation of the sphere.*

Connecting-rod crank system

Hero of Alexandria (?), 1st century BC; anon, China, 1st century BC; Watt, 1779–1781; Washbrough, 1779; Pickard, 1780

The connecting-rod crank system is one of the inventions fundamental to mechanics; indeed, it is indispensable for the translation of a **rectilinear movement** into a **circular movement** and, among other things, it has governed the exploitation and development of **steam engines**. A **crank** consists of an arm perpendicular to a shaft to which it can impart a circular movement, and a **connecting-rod** is a rigid bar which has at each end joints called **connecting-rod heads**. With the help of these heads the connecting-rod effects the communication between two mechanical organs, for example the end of a piston and the crank which transmits the movement to a wheel.

The origin of the invention, for it really is one, is controversial. In the 1st century BC. Hero of Alexandria spoke of a handle which was used to turn a shaft, but it has not been proved that this actually was a crank. The connecting-rod on the other hand features, at least in principle, in Ctesibius's **suction and force pump**. It was also found, though only in principle, in his final **organ** (see p. 70), for it is assumed that this had only one head; it was therefore used as a **lever** just as much as a connecting-rod. A specialist on the engineers in ancient history, Bertrand Gille, considers that 'the Greeks did not know the connecting-rod crank system', although he has published a reproduction of the **baroulkos** (see p. 101) which distinctly comprises a crank.

This point is all the more troubling when one considers that the **Archimedes screw**, which dates back to the 3rd century BC, was also activated using a crank (see p. 101). Although he knew both the connecting-rod and the crank, Hero of Alexandria and the other engineers of his time did not think of coupling them, or did not see a reason for doing so.

Otherwise its roots may have been in China. The crank was known there from the 1st century BC, as attested for certain by a clay sculpture made at the beginning of the Han dynasty, and reproduced by Robert K.G. Temple in his work on antique Chinese science. The invention probably dated to before the making of this sculpture (which confirms the possibility that Hero of Alexandria could have known about it from travellers, supposing that it was not from Archimedes's successors). In 530 nonetheless, a Chinese engineer, who unfortunately is unknown, was the first to put together the crank and the connecting-rod for the exploitation of energy, in this case, hydraulic energy; whence the description of a **hydraulic piston motor** which does include the connecting-rod crank system.

Nothing more was heard of it for many centuries. The only interesting derivative of the crank to appear subsequently in Europe was the **crankshaft**, which was included, for example, in Guido di Vigevano's 12th-century project of the crank-powered **propeller boat** and which prefigures the **camshaft**. In fact everything took place as though the connecting-rod crank system had never been invented.

In the 18th century, suddenly and in the space of two years, two men rediscovered the system. The cause of this awakening is obvious: the **advent of steam**. It was James Watt who was the first to reinvent it; in 1769 he thought of using two machines to simple effect which would activate the same wheel by the alternate action of two cranks; this was the **double-effect engine** (see p. 106). This posed the problem of the transformation of the alternating motion of the head of the beam, which is a type of crank with circular movement, and that of the connection of the stem of the piston to the other end of the beam. It is not known at exactly which moment Watt reinvented the connecting-rod crank system, but he is known to have used it in 1779, for that is the year when his unscrupulous compatriot Matthew Washbrough patented it in his own name; Washbrough had learnt about it from one of Watt's workers. In 1780 there was the striking coincidence of another inventor, James Pickard, also patenting a connecting-rod crank system. Hardly the sort to allow himself to become a failure, Watt patented an **epicycloidal gearing system**, the **planetary**, which was in fact a connecting-rod crank system designed to resolve the second problem of his double-effect engine, and which would lead to the **parallelogram** which bears his name.

Differential

Geminus (?), 87 BC

One of the most important mechanical inventions in history, the **differential**, seems to have been the work of a mechanic from Rhodes called Geminus in 87 BC. The invention and its date to within a few months are quite definite, and the story of the discovery is rather romantic. In 1900 some sponge fishermen located the wreck of a Greek ship lying 42 m under water off the island of Cythera, between Crete and the Peloponnese. The contents of the wreck were recovered a few months later; they included statues, amphoras and other objects which could be dated as going back to the 1st century BC. Among the fragments brought up to the surface, archaeologists found metal objects containing bronze cogwheels covered in thick layers of limestone and rust. These cogwheels aroused their curiosity because they were very finely made. After being cleaned they were to become the object of sporadic attempts at interpretation, which were all more or less fruitless. In 1957 the English scientist Derek de Solla Price proposed to carry out the first thorough analysis of them, which he concluded in 1975 and which agreed satisfactorily with the available relics. The cogwheels were part of a device which was made up of 13 cogwheels, of which 11 were placed on top, or if one prefers, 24 cogwheels with 11 superimposed, on a set of 13 axes. These wheels each possessed a definite number of teeth so that they exactly matched the **cycles of the Sun and the Moon**. The device evidently had no motor and must have been operated manually; for example, for a given day of a given month it could establish the moment of the Moon's cycle, even if it were hidden by clouds. Derek Price's reconstruction has been accepted by the whole of the international scientific community, despite some aspects of it being disconcerting.

It is disconcerting for two reasons. The first is that it reflects a knowledge of mechanics which would have allowed the invention of the first differential, that is, the first chain of gears using which one can transmit to a rotating shaft a movement equivalent to the sum or to the difference of two other movements. Though it is attributed theoretically to the Chinese (see p. 82), the invention of the differential is in no doubt here, because it is based on definite archaeological elements. The second reason why Price's reconstruction is disconcerting is the degree of evolution of Greek technology in the 1st century BC: contrary to all historical logic, this has actually enabled the creation of a wheel of about ten centimetres in diameter which comprises 27 teeth. This is a feat which was not achieved again until the advent of **clockmaking**, ten centuries later. The nature of this masterpiece and the reasons why it apparently had no sequel remain to be explained.

Price attributed the invention of the object, known under the name of 'Cythera's clock', to Geminus, an astronomer who was

> Geminus of Rhodes is not a major figure among the early Greek scientists, although his fame brought him Roman citizenship at the hands of Pompey and he was able to meet Julius Caesar. He had, however, a profoundly original spirit, being one of the first to have had the clear-cut intention of separating the knowlege about the world from philosophy, thus prefiguring science in the modern sense.

famous in ancient history for his mechanic's brain and who was the pupil of Posidonios of Apamea. The clock would have been built in Rhodes, the great astronomical centre.

The attribution of the invention to Geminus could cause problems. Indeed, Hero of Alexandria, who was far more famous than Geminus, mentioned the method of making **odometers**, one for land use, the other for sea use, in his treatise *Mechanics*. These instruments definitely included differentials; Hero would therefore be the inventor of the differential, and it does seem, moreover, that Geminus went to Alexandria. This is an interesting point, for if one examines the **reduction system** (see p. 101) also described by Hero, it will be noticed that it constitutes the stage preliminary to the creation of the differential. Whatever the case, Hero does not lose any credit if he is described as the teacher and therefore the inspiration behind Geminus's invention of the differential.

Electric generator

Guericke, 1663; Volta, 1775; Faraday, 1831; Pixii, 1832; Wheatstone, 1845; Pacinotti, 1860; Gramme, 1870; Edison, around 1878; Van de Poele, 1888; Lamme, 1896

Electrostatic forces were described for the first time by Thales of Miletus in the 7th century BC, but it was not until 1663 that Otto von Guericke, a teacher and philosopher in the German town of Magdebourg, who was interested in **pneumatic forces** had the idea of making an incredibly simple machine to produce **static electricity** (it is still not certain whether his invention was intentional). His invention could have been realized during any preceding century, for it was simply a ball of sulphur set on an axle which was turned using a crank; when both hands were placed around the ball it was excited electrically. This was the first **electric engine** of modern times; it was of the **'friction' type**. Isaac Newton made one out of a glass globe around the year 1709 and his fellow countryman Francis Hawksbee made a similar one the same year; he was the first to think of collecting the electricity using a metal chain which was coiled around the globe, like a driving belt. In 1787 the Englishman Edward Nairne replaced the ball with a glass cylinder mounted on feet which were also made of glass, and he was the first to collect **negative or positive electricity** as he chose. This type of machine had no practical use at all, but it constitutes a remarkable example of the irresistible attraction of the unknown.

In the meantime another type of electric machine appeared, which was more complex. It consisted of two **Leyden jars** or **accumulators** which were already charged with static electricity with brass balls A (positively charged) and B (negatively charged) on top; two other balls were then placed nearby, A' and B'. It followed that the hemisphere of A' near A would become negatively charged and the other hemisphere would become positively charged. The inverse phenomenon occurred with B', of which the hemisphere

The history of the electric generator is interspersed with the names of workers whose efforts have not been acknowledged, either because they did not go deeper into their research, or because they did not patent their discovery in time. This was the case with the Englishman Peter Barlow who, in 1823, eight years before Faraday, made an embryonic generator consisting of a studded wheel half immersed in a tub of electrified mercury which turned under the action of a current. This was also the case with the Frenchman Louis Clerk who invented the collector in the same year as Gramme and whose name is all but forgotten.

near B would be +, and the other, −. If for the sake of argument, the + hemisphere of A' was linked to the − hemisphere of B', they would cancel out and lose their electrical charge: A' would remain entirely negatively charged and B' positively charged. If the positions of A' and B' were swapped round, it would follow that the charges of their Leyden jars would increase. This is because the negative charge of A' would be added to that of B and the positive charge of B' to A, for the two jars, being connected to the ground, have a greater capacity than that of the balls. By carrying out the operation several times the charges in the jars could be increased. This is called the **'influence' type of machine**, where mechanical work produces electricity, and where the type of apparatus enables it to be collected. This device, the Italian Alessandro Volta's **electrophorus**, was the first **self-exciting dynamo**, as well as being the first in the long series of influence machines which existed long before the 20th century. The main ones were those of the Englishman Abraham Bennet (1787), his compatriots Erasmus Darwin and B. Wilson, the German G.C. Bohnenberger, the Frenchman J.C.E. Peclet, which range from 1788 to 1888, the Englishman C.F. Varley (1860), the German A.J.E. Toepler (1865), his compatriot W.T.B. Holtz, who built several between 1864 and 1880, the Englishman James Wimshurst (1878), and the German Robert Voss (1880). From the end of the century more and more interest was shown in machines of this type because they allowed a convenient **electrical induction** to be realized at a low cost. It was in this way that a crank was used to start the internal combustion engines in motor cars (until World War II) and also how the current was induced in telephone-stations.

In 1831 the Englishman Michael Faraday conducted an experiment comparable to that of Guericke, but not similar. By making a copper disc turn between the arms of a magnet, he obtained a continuous voltage at the edge and at the centre of the disc; he had discovered **electromagnetic induction**. He had also demonstrated that there was another way of transforming mechanical energy into electric energy apart from using the electrostatic machines. His instrument had only a mediocre output, however, and in 1832 it was somewhat modified by the Frenchman Hippolyte Pixii, on that occasion by adapting Ampère's **commutator** from alternating to direct current.

In 1845 the Englishman Charles Wheatstone replaced the permanent magnets with electromagnets which were fed with direct current by a battery, modifying his own machine seven years later in such a way that it became self-exciting: the coils of the magnets received directly the current which was produced by the rotation of the disc. It was only with the improvements made by the Italian Antonio Pacinotti (1860) and the Belgian Zénobe Théophile Gramme (1870), however, that Faraday's machine began to have an interesting output: Pacinotti and Gramme had the idea of putting small metal plates in contact with the turning part, or rotor, of the generator; they had just invented the **collector** (the credit for which seems usually to go entirely to Gramme). The trick had been essentially technological: in order to collect the current the predecessors of Pacinotti and Gramme had had recourse to shunts fixed to the armature of the rotor (in this case, coiled-up armatures). By installing the coils of armatures on the rotor itself, Pacinotti and Gramme reduced the volume of air in which the magnetic field developed; thus they strengthened it and consequently raised the electrical productivity of their generator. Only one improvement remained for them to add: given that the rotation of the armature itself generated alternating currents, not only in the wires but also in the steel supports, and that the **eddy currents** circulating in the steel caused losses of electricity, they laminated the supports.

From this time onwards the generator went far beyond the status of fragile laboratory apparatus which it had been until then; thenceforth it became a robust machine intended for industrial use, as witnessed by Gramme's machine which is kept in the Industrial Arts and Crafts Museum in Paris. Electricity losses happened all the same, and the Germans Ernst Werner von Siemens (1856) and then Friedrich von Hefner-Alteneck (1872) tried to mitigate them by modifying the coils in the electromagnets.

It was however the American Thomas Alva Edison who again reduced these losses in a spectacular way, by deciding to simplify one of Pacinotti and Gramme's ideas. These men had thought of raising the voltage of their generators by increasing the number of magnet poles between which the rotor functioned; Edison reduced these poles to two and in this way increased the output of his generator by 50 to 90%, which was unheard of (and moreover enabled Edison to build large electricity distribution networks). His first generator was ready in 1878. However much of a genius Edison was, he was a self-taught man and did not possess the knowledge which would have enabled him to establish the rational principle behind the making of a generator; the credit for this went to the English engineers John and Edward Hopkinson.

In 1881 the American Charles Brush thought of doubling the magnetic field coils. In this way he made the first **double magnetic field generator** or **compound generator**, which boasted a remarkable advantage over earlier generators but also a major problem. The advantage was that the voltage remained constant whatever the charges, and the problem was that the brushes on the corrector — which replaced Gramme's small plates — wore away very quickly, as did the rods of the commutator. The generators of the time produced sparks in an impressive way. In 1888 the American Charles J. Van de Poele invented carbon brushes, which solved the problem.

In 1884 the invention of the **electrical transformer** by the Frenchman Gaulard considerably changed first the research and then the electric installations. Thenceforth generators (once again multipolar) which were benefiting from the American Benjamin Garver Lamme's invention of the internal current **equalizers** (1896), became increasingly large and powerful. It became possible to produce a current from steam and hydraulic energy and to convey it a great distance. This improvement had immense repercussions, both economically and industrially; energy was available everywhere.

Escapement (in clockmaking)

Gerbert, around 950; Hooke, 1666; Graham, 1715; Le Roy, 1748; Mudge, around 1755; Berthoud, Arnold, Earnshaw, around 1780

Escapement is the mechanism which controls the revolutions of the main cogwheel in a **clock**, by assuring it the necessary regularity; it is therefore the essential component in clock making. The principle was invented by the Chinese I-Hsing (Yi Xing) in 725 and was used in the first known **mechanical clock** (see p. 152). Any escapement implies a force which makes the cogwheel turn; in I-Hsing's clock this was **hydraulic power**. In the first clocks to appear in the West, it was **gravity**: a weight was attached to the axle of the cogwheel by a string coiled around it; the unravelling of the string caused by the weight was slowed down by a pivoting **weight balance** at the centre of which an A cleat or anchor was fixed. Each arm of this cleat fitted alternately between the cogs of the wheel, letting a cog escape at a determined angular moment of its pivoting. The cleat therefore stopped the wheel for a given time, which corresponded to a fraction of an hour or a minute, depending on the length of time corresponding in turn to one whole revolution of the wheel. This mechanism is known under the name of **foliot**.

The exact date of the appearance of the first mechanical clock in the West is not

It is worth noting that the first complete theoretical account of the **function of the anchor escapement** was written only in 1959, by Professor R. Chaléat of Besançon.

known; although it was there that the first non-hydraulic mechanical clock was made. It is often said that the oldest one of which there is an account is that erected at Westminster in 1288; it seems likely however that mechanical anchor escapement clocks were made previously. With the principle of the foliot generally being attributed to Gerbert d'Aurillac, who became pope in 999, and the date of the invention being around 950, it can be supposed that the first foliot clocks actually date from the beginning of the 11th century. Among other reasons for claiming such a distant origin, there was Gerbert d'Aurillac's interest in clockmaking and the power he had when he wore the tiara under the name of Sylvester II, which would have enabled him to have his invention put into operation.

Balance-wheel escapement was terribly inaccurate, and could result in differences of up to one hour in twenty-four, as can be judged from the oldest existing specimens. Considerable progress was made with Galileo's discovery of the timekeeping property of the **pendulum** in 1581. The advantage of the pendulum is that its **period** is independent of the path it follows and depends only on its length (if one excepts the practical variations caused by the stretching of the material). It therefore offered a basis of regularity which was quickly seized by clockmakers. When it first appeared balance-wheel escapement was seen to include an integral inaccuracy, which resided in the determination of the delay of escapement by the anchor. A **free escapement** was therefore sought which benefited fully from the regularity of the pendulum. Innumerable projects came to light, the most striking of which was proposed by the Englishman Robert Hooke in 1666 and made a few years later by his compatriot William Clement. This invention consisted of placing the cogwheel and the anchor on a single plane, whereas in balance-wheel escapement, the anchor was perpendicular to the wheel. This invention was advantageous in that it reduced the inequalities, and its efficacy was undeniably increased by improved clock components manufacture in the 17th century, as well as by the progress made in the manufacture of hard alloys which enabled much more delicate work than was possible with the iron foliots of preceding centuries. It attained an unprecedented reliability in the form given to it in 1715 by the Englishman George Graham, which is known as **Graham** or **deadbeat escapement**. In 1725 Graham also invented **cylinder escapement** for use in watches. The cylinder in this was mounted directly on the axle of the spring-loaded balance wheel and, owing to its shape, fulfilled the function of **regulator** of the anchor.

In 1748 the Frenchman Pierre Le Roy invented the first real integral free escapement. The anchor received an impulse which it transmitted to the balance wheel — which lifted it to allow one tooth to escape. This happened only in the direction of the movement of the balance wheel, and not when the balance wheel recoiled in the opposite direction. In theory this new device aimed to do away with the constraint regarding the time-lag of escapement of the two ends of the anchor, a problem which Hooke had come up against and been unable to resolve. It was nearly perfect, for it worked on the basis of a single movement which was governed solely by the movement of the cogwheel, and it allowed the second movement to be lost; it is known as the **chronometer**. This principle, which was to need much improvement, was apparently used only in two 'marine watches', where it was hardly very satisfactory; it effectively operated like large **chronometers**, and the theory was not taken up again until around 1780 by the Frenchman Ferdinand Berthoud and the Englishmen John Arnold and Thomas Earnshaw, who modified it by devising a **single-beat escapement**. This had two great disadvantages which slowed down its development: it was cumbersome and did not work in all positions, which is why it was never used in watches. A really practical free escapement remained to be invented.

It was at this juncture that, in 1765, the Englishman Thomas Mudge — for it was in England that modern clock and watchmaking first began — had made a remarkable invention, which was reliable and which further reduced the **friction** caused whilst the parts were in use, for example in the Graham escapement. The anchor, which was of a design very similar to that of marine anchors, lifted as soon as the tooth of the wheel had transmitted its impulse; the anchor transmitted the impulse to the balance wheel which lifted it up in order to let the cog escape, then the next cog was blocked on its stop; in the meantime, another cog had begun its course and had just transmitted another impulse, and so on. This system necessitated a certain amount of space between the cogs. It reduced the friction by blocking the wheel between the escapement of one tooth and the next, instead of maintaining the constant sliding of the ends of the anchor, called slopes or surfaces of impulsion, on the teeth. This is the **lever escapement**.

Since then numerous types of escapement have been invented which more or less faithfully follow the principles explained above.

Float regulator

Hero of Alexandria, 1st century BC

One of the most remarkable examples of **automation** (see p. 82) in ancient mechanics was the float regulator. It was invented by Hero of Alexandria, based on the principle of **communicating vessels**, and was intended to indicate the filling level of a container. It consisted of two containers, one of which poured the liquid and the other of which received the poured liquid. These two containers were linked by a tube which did not empty directly into the second container but into a small receptacle which was placed inside it and which floated on the surface with the help of a cork support. On the side of this small container there was a lip situated at a certain height. The liquid poured by the first vessel flowed out freely, first into the small container with the float and thence into the second container, until its water-level was the same as that of the smaller one. The difference between the water-level of the small container with the float and that of the larger container in which it floated was defined by the **water-line** of the small container, and this determined the desired filling level of the second container. This was called the **floating siphon regulator**; it made use of the pressure of the air and was in fact a precursor of the **barometer**.

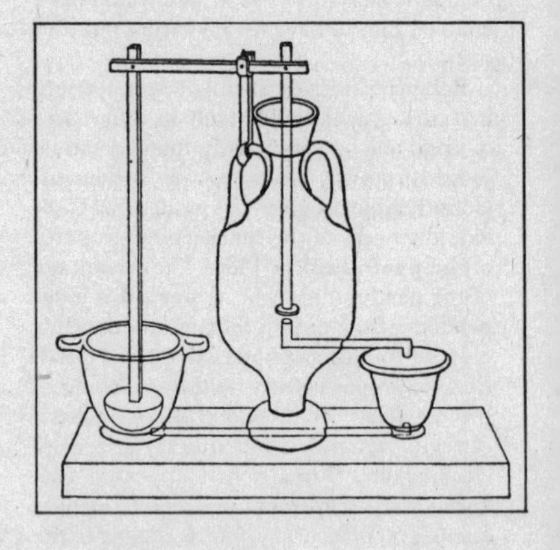

A float regulator

Hero also invented a second type of regulator, called the **weight regulator**. In this the container to be filled was balanced with a predetermined weight, which corresponded to the sum of a certain quantity of liquid and the weight of the receptacle. When this was full it lowered the beam of the balance on which it was placed; at the same time the weight rose and lowered a rod which closed the siphon through which the liquid was being poured. This system was intended for the automatic distribution of determined quantities of wine and it is in fact the ancestor of the **optics** in modern bars.

Floating mills

Belisarius (?), 537

Floating mills were a widespread source of energy until the Middle Ages, and then they were mysteriously forgotten. They were derived from proper **water-mills** and were a half-way stage in the development of the **turbine**, subsequently being called **mill-boats**. They consisted of sets of several vertical wheels mounted on the same shaft on both sides of a boat or landing stage which was well anchored in a water current, and they powered **mills** by means of a **reverse gearing** system. They were placed on rivers at points where the current was particularly strong. The invention is attributed to the Byzantine general Belisarius who was besieged in Rome by the Ostrogoths in 537. Since the aqueducts which provided the water for the town's flour mills had been cut off, Belisarius or his engineers had the idea of taking the gearing systems and the mills to the Tiber and working them with sets of **paddle-wheels**. The Byzantines' inspiration may have been drawn from the Near East.

The floating mills had lasting success: in the 14th century there were 68 of them on the Seine, used to power **oil presses** and **stone-cutters**; they also featured on the lower Rhône and on the Rhine, where they operated **paper mills**; on the Elbe and the upper Danube, where they provided the energy for **sawmills**; on the Adige, where they powered **mortars** for the manufacture of **gunpowder**, as well as on all European rivers. Some still existed on the Tiber in the 19th century and they are still used in Yugoslavia (during World War II the Germans used them as flour mills on the Danube).

Fuel oil (energy-providing use of)

Anon, China, 4th century BC

Natural hydrocarbons, particularly **bitumen** and the more liquid **petroleum naphtha**, have been known since ancient times, especially in the Middle East where they are plentiful. It was in the province of Szechwan in China during the 4th century BC that they were first used to provide energy. They were used for cooking food, burning in lamps, making torches and blasting rocks in public works (the rock was covered in bitumen which was ignited to make it blow up).

The first uses of bitumen and its derivatives in Europe seem to have included the **lubrication** of chariot axles and the preparation of **medicinal ointments**.

Heat pump

Kelvin, 1851

The principle of the heat pump existed nearly a century before the first models were made; it was conceived in 1851 by William Thomson, Lord Kelvin. It consists essentially of an inverted refrigerator, which produces coldness using an evaporator and heat using a condenser. In the heat pump or **thermopump**, the heat provided by the evaporator is collected using a compressor so that it can be used for heating.

It was around the middle of the 20th century that reversible air-conditioning systems began to be made. These used valves which reversed the circulation of air.

Hydraulic pumps

Archimedes, 3rd century BC; Ctesibius, 3rd century BC; Hero of Alexandria, 1st century BC

The making of mechanisms which could convey water from one level to another and then draw it up or invert its flow reached a remarkable degree of sophistication and efficacy at a very early stage, by at least the 3rd century before Christ. The most famous of these mechanisms is attributed to Archimedes: the **screw** which was named after him, and the construction of which was already quite complex. It consisted of seven partitions (in the Greek and Roman models) which were fixed in a spiral on a log in such a way as to create the same number of compartments. This was then covered with a cylinder for a streamlining effect and made watertight with coal-tar, with only the two ends remaining open. Two iron axles were driven into each end of the log and placed on a fork, and one of these was equipped with a handle enabling it to be turned. The appliance was submerged in water, such as into a river, at an angle which the Roman architect Vitruvius recommended should not exceed 37°. When it was rotated using the handle at the top end the water could be channelled through the internal spiral compartments up to the higher level. The Archimedes screw was very commonly used in agriculture for the irrigation of land which was not much higher than a river or a canal. When operated by hand an appliance such as this, if it measured 2.4 m and had a diameter of 0.3 m, could produce the considerable flow of water of 160 to 235 l a minute.

Some information left by Vitruvius would seem to suggest that this type of **rotor** was not powered by the arms but by the legs, but any information about a mechanism which enabled the Archimedes screw to be driven by the legs has been lost.

Another type of pump used in ancient times was the **tympanon**. This consisted of two **drums** which were linked by eight partitions with cusps on their outside edges and which turned on an axle. Around the centre of just one disc there were eight openings, one in each compartment; below the centre there was a parallel separate gutter. The tympanon worked in the following way: it was submerged in the flow of water

Ancient pumps were cast in bronze. The reaming seems to have been achieved by force, using olive oil as a lubricant (although it has little mechanical virtue). One of Vitruvius's reports suggests that either for the reaming or during operation, the pistons were covered with untanned sheepskin.

to a depth of half or a third of its height, and it collected the water which was captured by the cusps. The water flowed out through the central openings into the gutter as the mechanism turned. As with the Archimedean screw, the tympanon allowed water at one level to be collected and conveyed to a higher level. A tympanon of 3 m in diameter and 0.2 m thick would have provided 400 l of water with one turn and 800 l in a minute.

Vitruvius also described a system of buckets attached to a chain which was moved by using a winch with a handle, the principle of which was derived from the tympanon. It was however with the **pneumatic organ** made by Ctesibius, an engineer from the school of Alexandria, that the foundations for modern pumps were laid. The principle behind it was the following: two cylinders with a hole pierced on their lower surfaces were equipped with **pistons** activated by rods fixed on to a **balancing rod**. The cylinders were connected by a central horizontal pipe to which another pipe, the drainage pipe, was connected. When a piston descended and compressed the air below it, it created a pressure which injected the water towards the drainage pipe, and its lower opening, consisting of a floating disc, closed because of the pressure. Meanwhile, the other piston sucked in water because of the perfect piston-cylinder watertightness, and when in turn it was lowered, it too pushed the water back towards the drainage pipe. This was the first known model of the **suction and force pump** which could function continuously.

The pneumatic organ was used in ships and when it was mounted on wheels above a water tank it acted as a **fire pump**. It was also used by the council services in numerous towns throughout the Roman Empire, and in irrigation systems.

The organ is particularly interesting for two reasons: firstly, because it implies remarkable mastery of the **reaming** of the cylinders; secondly, because the rods of the pistons were mounted on **pivots**, in order to prevent the lateral movements of the pistons.

Hero of Alexandria, another technological genius of ancient times, improved Ctesibius's principle by providing the drainage pipe with a head which could pivot 360°, allowing the jet to be directed in any desired direction, and by making **disc valves** for the input of water, thus reinforcing the watertightness of the cylinders.

Hydraulic ram

Montgolfier, 1796

The hydraulic ram is both a **lifting machine** and a **pump** of an extremely original kind, for it needs only the energy of a waterfall and requires practically no maintenance. It was invented in 1796 by Joseph Montgolfier, one of the two brothers who made the aerostat which was named after them. The principle behind it is as follows: the falling water is directed into pipes which comprise a first valve or impulse valve A. Then it follows its course up to an ascending bend in the pipes, where there is a second valve B; this runs into a watertight reservoir. At the base of this reservoir there is an opening into another pipe, also with a bend in it, and with an ascending part leading to a second reservoir which is much higher than the upper level of the falling water. When the water falls into the first set of pipes, valve A is open; the water goes back up towards valve B, following the principle of communicating vessels, and pushes it shut, because it compresses the air which is at the top of the reservoir. The speed and pressure of the liquid then increase rapidly and suddenly shut valve A. This same increase in pressure forces valve B to reopen, immediately raising the air pressure, which sends some of the water from the first reservoir towards the storage reservoir, and then closes valve B again. In the meantime valve A has re-

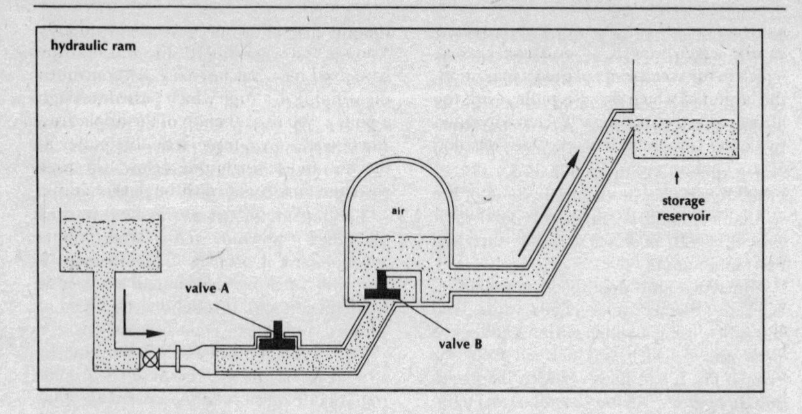

opened and the cycle recommences.

The hydraulic ram, thus named because of the 'butting of the ram' and because of the impact of valve B when it closes, can function with a minimum fall of water of 1 m and raise it to 100 m. Though seldom cited in technological histories, it was reintroduced successfully in France during World War II because of energy shortages, and used for the irrigation of farmland.

Lifting appliances

Hero of Alexandria, 1st century BC; Kyeser, 15th century; di Giorgio Martini, 1470–1480; Leonardo da Vinci, c.1500; Golovin, 1770

Lifting machines, which are essential to building, works of art, trade and industry are as old as architecture itself. The inventions which constitute their history defy inventory, for during the centuries between the birth of these techniques and the Industrial Revolution there were as many inventors as there were towns, and each inventor may have proposed several inventions in succession, particularly of lifting instruments, with each one being adapted to a particular purpose. The embryo of the first lifting devices was incontestably the **pulley** which, in reducing the friction of the rope on the transverse lifting axis, also reduced the effort required from the men or animals. Some theoreticians have speculated that the Egyptians used a lifting appliance called a '**goat**' for the building of the pyramids. The goat would have been a portico, free at the base, with a crosspiece which would have served as a support axis for a rope, shorter than the length of the uprights, and to which the weight to be lifted would have been attached. The weight would have been fixed to the rope with the portico in an inclined position. The portico would be wedged at the base, and raising it by traction would have carried the weight above its wedged point, to a distance equal to that from the wedge to the loading point. It is an attractive hypothesis, but doubtful with regard to the strength of the vegetable ropes of the time.

If speculations are abandoned, the first lifting appliance which can be dated and attributed with reasonable certainty is the **baroulkos** which was invented and made by Hero of Alexandria in the 1st century BC, and which is described in his work *Mechanics*. This device, which is also credited with being the first known **reduction**

system (see p. 101), was operated using a **crank** which moved an **endless screw**, which in turn set a series of **gears** in motion, the largest of which drove a pulley with the lifting cable wound on to it. Hero invented two other lifting appliances. One consisted of an endless screw with a large thread which was activated by a **capstan** situated in a vertical axis. At each turn this pulled a rectangular piece or wedge of wood, which was lodged between the thread of the screw and a groove in the rectangular frame which encased the whole thing. The other also consisted of an endless screw activated by a capstan which set going a **cogwheel**, the axle of which caused the winding of the lifting cable. The first of these appliances makes one a little sceptical, for it necessitates a minute adjustment of the wedge of wood to the grooves of the endless screw, and anyway its functioning comprised considerable friction. The second seems more feasible and reveals that manufacturing techniques of screws and cogwheels were already very advanced.

Hero's inventions are interesting because they introduce mechanical principles which do not seem to have existed before his time. The reduction system in particular is a form of **transformation** of a **circular movement** into a **rectilinear movement** which for the first time enabled **constant traction** to be applied to an object which was too heavy to be lifted or forced directly by an agent, animal or human, and which could move it by different means. When studying an invention derived from the baroulkos, and which is in fact the first known **crane**, the English engineer J.G. Landels calculated that a healthy worker could lift a 2 t block using this appliance.

The date and place when the crane first appeared are not known, but it seems to be a version of the goat described above, enriched by the observation of Hero's baroulkos, unless the opposite is the case and the baroulkos is an improved version of the cranes which may have existed previously (and which could have been used for the erection of monuments such as the Parthenon in the 5th century). The fact remains that the first descriptions available to us were left by the Roman architect Vitruvius from the 1st century before Christ. These cranes consisted of two uprights in the shape of an A, evidently made of thick wood, with the width at the base fixed by a capstan axis; this controlled the winding of a rope which passed through a pulley system at the top of the apparatus. There was, for example, a double pulley at the top and a single one below; the rope was therefore threaded through three times and thus was strengthened. This type of crane was called the **triple-pull** or **trispaston**, but there were also systems with five pulleys which were called **pentaspaston**. Some of these cranes had a set of hooks at the end of the lifting rope which were described by Hero as '**crabs**', though it is not known how they were closed. One version of the crane possessed a reduction system: the rope which was coiled around the capstan brought a drum of variable diameter into play around which it also coiled; this drum was mounted on a second axle which itself had two ropes wound round it which were used for the traction. This apparent complication, which caused a loss of energy through friction, had the advantage of being able to distribute the traction exerted on the rope better.

Some **cranes with a beam on a pivot** which was placed at the top of the A were apparently, according to the Greek historian Polybius, used as machines of war, either for throwing stones from the top of the ramparts of towns under siege, or in the employment of a curious military technique to overturn ships. In this latter case a diver would attach a hook to the prow of a ship; a rope attached to this hook would lead to the crane. Several men exerting a strong force of traction then pulled the prow of the vessel out of the water, which would take in water by the stern and capsize.

Perhaps there were later inventions of lifting appliances; Gothic cathedrals in any case make one think that they were in extensive use when they were built. Only one invention, however, has come to us since then, dating from the beginning of the 15th century, for it appears in the manuscripts of the German Konrad Kyeser which date from around 1430. It consisted of a **double effect elevatory appliance**, in fact a **double crane** activated by a single winch. Around the winch were wound two ropes which were carried by two opposing cranes and which each passed through three pulleys. The device was undoubtedly invented by Kyeser. Between 1470 and 1480 the Italian Francesco di Giorgio Martini devised a reasonably complicated appliance specifically for lifting columns. It consisted of a frame inside which two endless screws of opposing directions were fixed. The screws were activated by capstans, and they also moved two other linked frames using a rope. As the first frame went down because of the traction of the rope it made the second one move, which was placed on top and pivoted between two vertical uprights. The end of the second frame was lengthened by a structure with free beams attached to a hoop, and raised the top of a column resting in this hoop. The column was raised in stages, with the way to each higher stage being locked by a track system which blocked the upper frame. The advantage of this lifting system, a refined version of the Roman cranes, lay in the tracked blocking. This prevented the type of accident caused by the failure of the purchase.

New progress was made around 1500 by Leonardo da Vinci: the first **pivoting crane**. Unlike previous cranes, this was not limited to movements up and down and from front to back, but could carry a load laterally. It was placed on a base covered with a circular platform, on which the actual lifting apparatus was fixed. This was also the first known model of the **jib-crane**.

In 1770 the Russian F.D. Golovin improved the rotating jib-crane by inventing **rectilinear translation on rails** for the loading and lifting of the loads. A short time later cranes with **cast-iron bases** appeared, then those made entirely of metal. The advent of steam was to increase the power and diversification of lifting appliances considerably.

Natural gas (use of)

Anon, China, 4th century BC

The presence of numerous pockets of natural gas or **methane** on the surface of the soil and their eventual spontaneous combustion must have given the people of the southern provinces of China, especially Szechwan, the idea of making use of it, and even, what is more remarkable, of exploiting it. A text by Ch'ang Ch'ü (Chang Qu) dating from 347 clearly describes the construction of bamboo pipes, which were caulked in **bitumen**, which served to transport this methane to the towns where it was used for **town lighting**, among other things. This was undoubtedly the first example of **gas lighting**. More remarkable still is the fact that this methane was stored in bamboo tubes and used by travellers as torches or fuel reserves. In the 1st century BC the Chinese bored the earth systematically to collect methane.

It is assumed that the consumption of methane must have caused quite a few accidents, for even though the gas found on the surface burns without danger, the gas which is collected at great depths is much richer and has to be mixed with air before it can be used, to avoid explosion.

Reduction system

Philo of Byzantium, 3rd century BC; Hero of Alexandria, 1st century BC

The principle of the reduction system is that of the transmission of movement between two cogwheels of different diameters causing a change in speed and power. If the movement is caused by the smaller cogwheel, the movement of the larger one will be slower, but its force greater, and vice versa. This fundamental principle in mechanics — which is commonly used in the 20th century in **motor car gear-boxes** — seems to have been described for the first time in the 3rd century BC by the Greek mechanic Philo of Byzantium in a treatise on the **traction of heavy weights**, of which we unfortunately have only indirect knowledge. The levering device which he described is identified under its Greek name **baroulkos**, and we know about it through the version taken up by Hero of Alexandria in the 1st century BC in a treatise on mechanics which has survived. The baroulkos looked like a metal box inside which a screw thread activated by a crank made a series of progressively larger and larger gears move, the last one being a winch on to which the lifting rope was wound. The baroulkos therefore enabled an increase in power in proportion to the slowing down of the movement. It is likely that it was derived from a previous invention, the **screw elevator**, where an endless screw activated by a capstan moved a cogwheel attached to the lifting winch, a device also described by Hero, but of which he is not definitely the inventor. It is possible too that the principle of reduction was conceived in the 3rd century by Archimedes who had already noticed in the gearing systems that 'the large circles prevail over the small ones'.

The importance of civil engineering work, which was growing with the spread of the Roman Empire, ensured that lifting devices such as the baroulkos became widespread for use in the handling of heavy objects. Vitruvius made great mention of it in a treatise on agriculture in the 1st century.

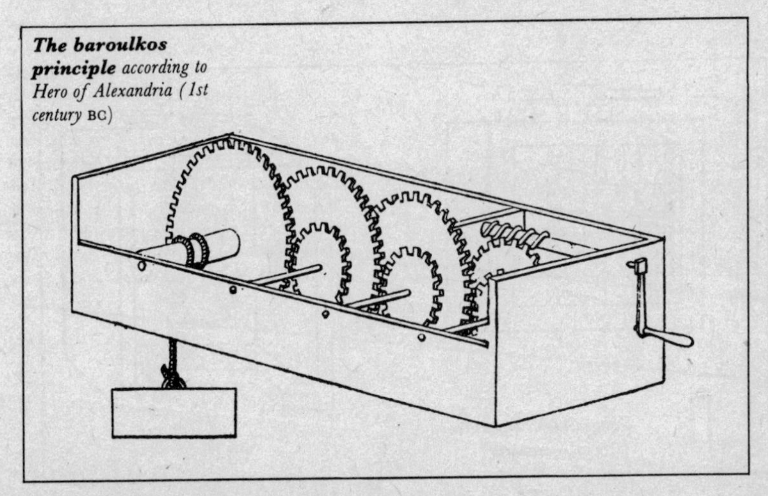

The baroulkos principle *according to Hero of Alexandria (1st century* BC*)*

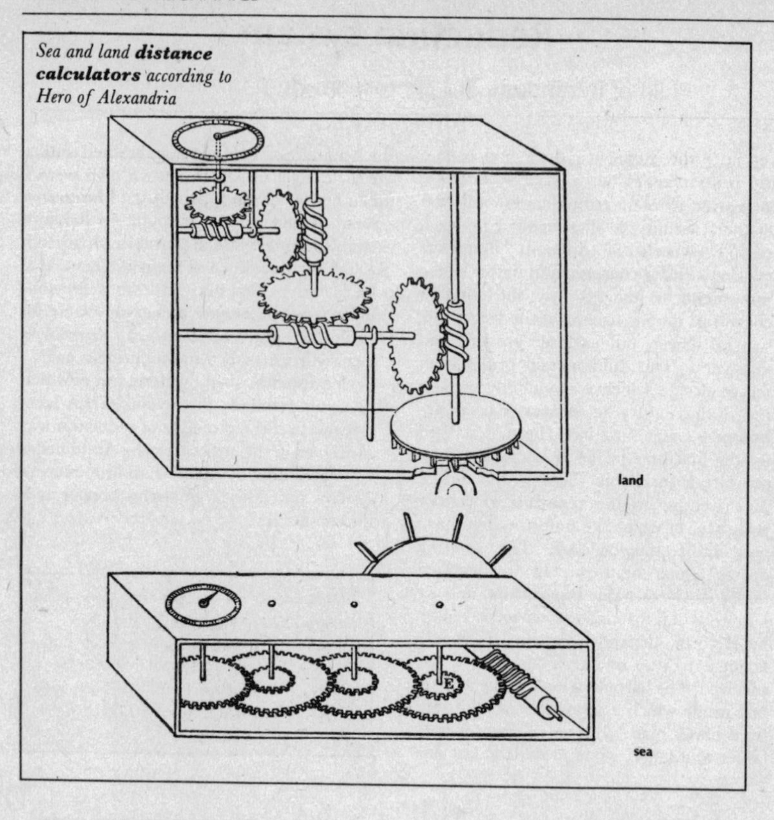

Sea and land ***distance calculators*** *according to Hero of Alexandria*

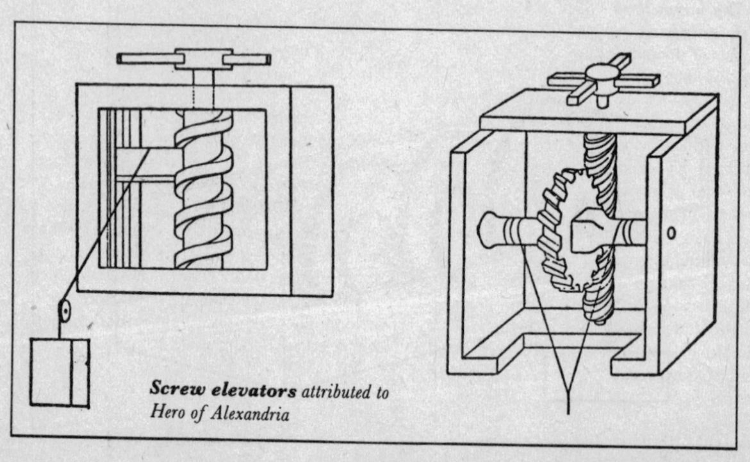

Screw elevators *attributed to Hero of Alexandria*

Saw

Anon, Egypt, c.4000 BC

Ever since the Stone Age there had been **flint instruments with jagged edges** which were fixed on to shafts of wood or bone, and used for cutting skins, chopping logs of wood or for harvesting. However their V-shaped heads obviously did not allow an even cut to be made when cutting thick materials. This was not obtained until the metal age; the oldest known saws were made first of **copper** and then of **bronze** around 4000 years BC in Egypt. Improvements in their length, sharpness and durability progressively enabled longer and more even cuts to be made. In 1500 BC the Egyptians succeeded in sawing up tree trunks into **planks**. From the mid-7th century BC, iron saws meant that they could saw stone, as attested by the marks found on the blocks of the pyramids. In the 1st century they used an invention of unknown date of which Pliny the Elder gave an account: the teeth were inclined alternately on both sides, thus widening the cutting edge and also creating a way of removing the dust created during sawing from the cut, and therefore reducing the amount of effort required.

One of the most paradoxical aspects of the history of technology is that after the fall of the Roman Empire, that is, after the 5th century, the saw nearly disappeared from woodcutters' and carpenters' equipment and was replaced by the axe. The analysis of the Bayeux Tapestry, which was made in 1100, indicates that William the Conqueror's fleet was made without using a saw. This instrument did not reappear until the 13th century.

Solar oven

Archimedes, 3rd century BC; Pifre, 1882

One of the most disconcerting breaks in the history of inventions concerns the use of **solar energy**. Its calorific power has been known since ancient times, and as far as we know the first to use it was Archimedes. In the 3rd century BC this famous inventor built giant **parabolic mirrors**, which he might perhaps have used to set fire to the Roman ships which were besieging Syracuse by concentrating a beam of sunlight on them. It seems likely that Archimedes would have made these mirrors not out of glass, but out of polished bronze; it is rather less likely that he succeeded in setting fire to the Roman vessels with them. The fact remains that it had been proved possible to capture solar energy. This remained dormant, however, until the end of the 19th century.

In 1882 the Frenchman Abel Pifre demonstrated the possibility of using solar energy in the Tuileries Garden in Paris. From 1 pm to 5.30 pm a concave mirror of 3.5 m in diameter, in the centre of which there was a boiler with a valve, activated a small 3/5 hp motor. This continuously operated a Marinoni press which was printing a newspaper written by Pifre, the *Sun-newspaper*, at the rate of 500 copies an hour.

Pifre had previously established that it was possible in the same way to heat 50 l of water to boiling point in less than 50 minutes, with the pressure inside the boiler increasing at the rate of one atmosphere every eight minutes. The demonstration remained without sequel.

***Solar motor** by A. Pifre, French engineer, dated 6 August, 1882. (Engraving by Poyet.)*
Concave mirror of 3.5 m diameter works a cylindrical steam boiler which itself operates a printing press

Steam engine

Hero of Alexandria, 1st century BC

One of the most astonishing gaps in the history of science and technology comprises the eighteen centuries which elapsed between the first steam engine and the discovery, combined with the application, of the principle of the **expansion of water vapour**, which was to inaugurate the Industrial Revolution and change history.

It is thought to have been in the 1st century BC that Hero of Alexandria, geometrist, inventor and creator of the famous Hero's formula, designed and built the first steam engine. It consisted of an extremely simple apparatus, called the 'aeolipile': this was a sphere with two spouts bent in opposite directions, pivoting on a hollow axle joined to a **boiler** which it filled with water vapour. The steam escaped via the angled spouts and made them turn. The device does not seem to have had any practical application at all; it was just a physicist's curiosity. Some theoreticians postulate that its principle could have been used in the secret and 'magic' machinery in certain temples which animated the jointed statues (which were also manipulated by counter-

It was only around the mid-20th century and as a result of the studies of historians such as Bertrand Gille that anyone realized the magnitude of the genius of Hero of Alexandria, one of the most prodigious mathematicians, geometricians, physicists, mechanics and inventors of all time.

Aeolipile
Greek manuscript of Hero of Alexandria, copied and drawn by Ange Vergèce (16th century)

weight systems and hydraulic mechanisms). The invention, which is described and sketched in *Pneumatica*, the two-volume work by the famous inventor, does not seem to have been taken up again before 1640 (1636 according to some); with regard to historical scruples, Leonardo da Vinci's research in this domain undoubtedly must be mentioned, although it does not seem to have led to any concrete realizations. The first is simply suggested by a rough sketch in the *Codex Leicester* apparently dating from 1504, which shows a receptacle of water below which there is a fire. There is a calf skin placed on the surface of the water which serves to make the container watertight, and on top there is a lid. The weight of this is balanced by a weight to which the lid is attached by two return pulleys. This would appear to be a sketch of a steam engine, but, as indicated by the appended notes, it is not: it is a device for measuring the expansion of steam. Little more convincing is another invention, apparently dating from 1509, the paternity of which is strangely attributed by Leonardo to Archimedes (probably to give it weight): it consisted of a **cannon** which was heated from above with a fire, and a **compression chamber** into which water was poured and turned to steam. The effect of the suddenly released steam would send out a bullet of a talent (26 kg) to six stages (from 887 to 1152 m). This steam cannon, which Leonardo called **architonitro**, is treated with scepticism (see p. 217). A third little sketch, still from the same source (the *Codex Leicester*), was intended to represent a steam **weight-lifting machine**: the lid of the heated container would lift a weight as it rose up. The drawing is too rudimentary for it to be objectively possible to attribute a reinvention of the steam engine to Leonardo with any degree of certainty, however.

It was around 1640 that the Marquis of Worcester turned to the expansion of steam to move a wheel with blades on fixed axles. However, whereas Hero used the steam in continuum, Worcester used it in bursts. Worcester had only three known predecessors, apart from Hero — Giambattista della Porta, Salomon de Caus and Giovanni Branca — who all stopped at the purely theoretical stage of invention.

Denis Papin also began from theory and for a long time he was held as the inventor of the steam engine. In 1687 he published a treatise, *Description and use of the new machine for raising water*, in which he described a machine functioning by *piston* under the effect of steam. The principle behind it was very simple: a piston sliding into a vertical cylinder, at the base of which the water to be heated had been placed, rose up under the effect of the steam; when the steam condensed the piston went down again and lifted the weight by means of a pulley. This was one of the very first **atmospheric engines** (see p. 81). Papin is ascribed with the application of this principle to **river transport**: it could have been used in a **paddle-wheel boat** built in 1707. This was not the case, however: the boat, which did exist, was propelled by manpower. Besides, it was destroyed by the boatmen of Hanover who were alarmed by the potential competition of this machine. In fact, Papin did not build any actual steam engine: the first real application of the principle outlined in his treatise was that by the Englishman Thomas Savery which dates from 1698 (Savery was the first to submit a patent on the steam engine). It consisted of a **mono-cylinder suction pump**, which drew water from wells and dried out flooded mines; like the one described by Papin, it combined the use of water vapour and atmospheric pressure (see p. 81). Though it was improved in 1705 by Thomas Newcomen, it remained slow, noisy, expensive and not very practical, and it was only in 1769 that it was significantly improved, thanks to the Scot James Watt's idea of installing a **compression chamber** on either side of the cylinder, and a **compartment** in which the vacuum, which was inherent in the ascent of the piston, could form. Watt himself was later to improve his own machine by using alternately the opposing forces of the steam and the vacuum.

All these machines were far from inaugurating the age of mechanical energy in the Western world. Even when the American Charles Avery built the first **steam turbines** for commercial use in 1831, many people were still sceptical of mechanical energy. It is worth noting, strangely enough, that the Avery machine was in a way a direct extrapolation of Hero of Alexandria's aeolipile: it consisted of a hollow axle which was joined to two tubes fixed at right angles. The steam escaping from these tubes made the axle turn, to which circular saws could be fixed. About fifty of these turbines were actually installed in sawmills (and one, experimentally, on an engine). Since they were noisy, quickly broke down and had speeds which were hard to control, they were abandoned. Half a century later a significantly improved version of them appeared, due to the Swede Carl G.P. De Laval, which achieved the record speed of 42 000 revolutions a minute. In 1897 De Laval credited himself with the invention of **divergent exhaust turbines** of powers ranging from 5 hp to several hundred hp. From 1829 onwards, in principle the steam engine created and conquered the railways.

Stirling engine

Stirling, 1816

Seven years before Sadi Carnot published his prophetic *Réflexions sur la puissance motrice du feu* (Report on the driving force of heat), which founded **thermodynamics** as well as the principles of an ideal reciprocating engine cycle, the Scot Robert Stirling, a clergyman like many of the scientists in Scotland, patented a heat engine, the true originality of which was not to be recognized until more than a century and a half later. Whilst the description of a cylinder and piston engine refers to a **single reciprocating cycle**, the operation of which includes the distinct stages of **expansion** and **compression**, the Stirling engine has **continuous operation**. This operation is based on a concept which is unique in the history of engines since Hero of Alexandria's aeolipile (see p. 105). The cylinder has not one but two pistons facing each other. The essential detail is that they are separated by a floating section of porous material in the cylinder: this acts as a **heat exchanger** or **accumulator** which absorbs a given percentage of the heat caused by the compression and therefore serves also for **storage** and **cooling**. First of all piston A compresses the air and raises its temperature, with the other piston B remaining at a standstill; then in turn piston B rises and pushes the compressed air through the regenerator; then piston A is pushed back by the air expanding in the expansion chamber. It would normally be cooled by the expansion, if it were not for the heat exchanger providing it with more heat. The expansion pushes cylinder B back to neutral, cylinder A goes back down again, and the cycle recommences.

An engine such as this is peculiar in three ways. Firstly it is an **external combustion** engine, with the continuously burning fire being situated outside the cylinder. It is an engine which can be installed in a coal boiler, for example, and it was first used industrially although this usage has diminished. Secondly, it functions using air and does not itself produce any exhaust gases, apart from those from the burning of the fire; it can therefore be installed in an electric boiler. Finally, its energy loss is very low because of the heat exchanger.

It nevertheless presented manufacturing problems which were so complex that the technicians who were satisfied with the four- and two-stroke engines quickly became discouraged. It was only in 1970 that engineers, in this case from the Dutch firm Philips, recommenced research on this engine which had for a long time been forgotten. They were attracted by the **absence of exhaust fumes**, which is a valuable quality in submarines for example. In fact the French civil submarine *Saga* is equipped with two Stirling engines which were made in 1985 by the Swedish firm Kockums. The engines are fed by pure oxygen.

The history of the Stirling engine evokes that of the steam engine in several ways. Although it was economical and non-pollutant it nevertheless suffered from a degree of unpopularity which in retrospect seems as unjustifiable as it is incomprehensible. Most histories of **mechanics** and **thermodynamics** published before 1970 do not even mention the Stirling engine or else they afford it only a few lines. Manufactured since 1818, the Stirling engines appeared to be 'completely outdated' by 1922, and no one took any interest in them any more. The single exception was the Dutch firm Philips who, in association with the American firm General Motors, took up the study of this machine and even built some 5000 hp engines. During the 1980s, nevertheless, numerous specialists reckoned that this archaic 'curiosity' had hardly begun its career. It is worth noting that this **isothermic** engine, which operates at constant temperature, has the same historical ancestor as the **diesel engine**: the **porous piston engine**. This acted as heat exchanger, and was built by the Englishman James Joule in 1884.

Suction and force pump

Ctesibius, 3rd century BC

The oldest known reference to the suction and force pump principle dates from the 3rd century BC; it is undoubtedly an invention and Ctesibius is accepted as the inventor. In a general history of technology it would be placed between the **mono-cylindrical hydraulic piston motor** and the **steam engine** (see p. 100). Also known as '**Ctesibius's machine**' or '**ctesibian machine**', it consists of two cylinders A and B, the bases of which are covered and submerged in water. They are joined by a horizontal pipe. Other pipes fixed at about a third of the way from the base lead up to a tank, the height of which varies according to the power of the pump. Cylinder A contains piston A' and cylinder B contains piston B'. The pistons are activated by rods which are joined by pivots to a single lever arm. When the lever is lowered so as to lower piston B', a force is exerted on the water which makes it go up into the corresponding pipe. At the same time, piston A' rises and sucks water into cylinder A. When piston B' goes up again and piston A' goes down, the water which was sucked

> Like Geminus, the illustrious inventor of Cythera's clock, Ctesibius was of humble origin: they were both sons of barbers. They both also acquired reputations which went beyond the bounds of their home towns.

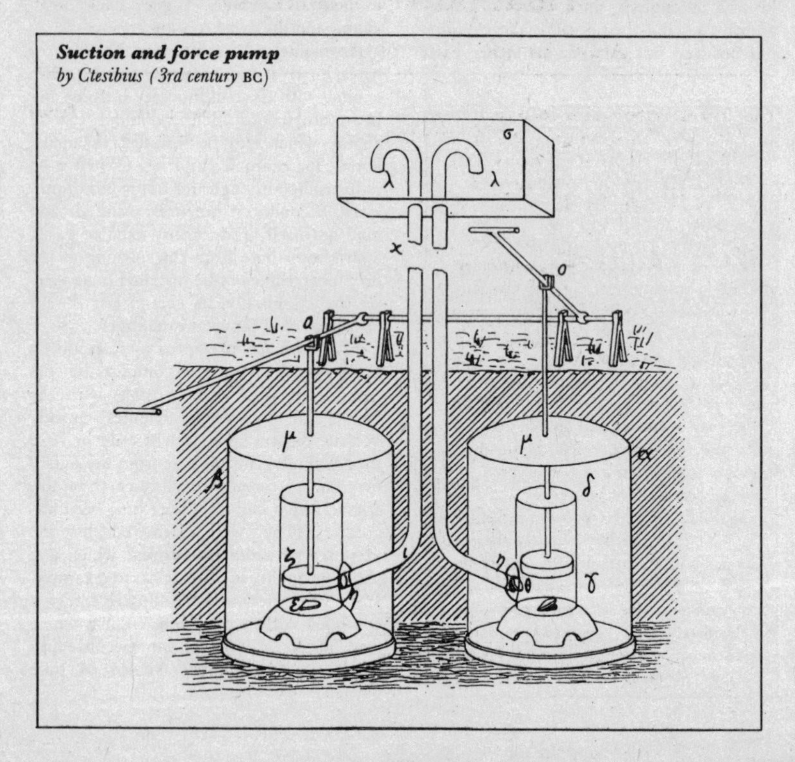

Suction and force pump
by Ctesibius (3rd century BC)

into cylinder A is driven towards cylinder B. The result of each of these coupled actions is that the water sucked in at an lower level is sent up to the upper reservoir. The pump can be used for sucking water, for example for draining a cellar, or for raising water from a lower level to a higher level. It was also used for animal-drawn **fire pumps**.

The manufacture of the suction and force pump must have met with technical problems, the main one being the **reaming** of the cylinders, a fundamental condition of watertightness, which is essential to the efficacy of the pump. The cylinders and pistons are known to have been made of bronze and lubricated with olive oil (which is not a very suitable mechanical lubricant). It is also known that a great many models of the instrument were made, for many have been found, in England and in Italy for example. Ancient technology must have been rather more refined than previously supposed.

The origin of the conception of this remarkable invention remains vague. It could have been derived from the **Chinese monocylindrical pump** (for there was trade between Asia and the Mediterranean). Or perhaps it came from the idea of a much older and widespread instrument, the **bellows**, that is, the expulsion of a fluid to a distance by compressing the contents. It was definitely an invention and not a discovery. It served as a basis for Ctesibius to make the first known **organ** (see p. 70).

Tidal power stations

Anon, Europe, 11th century; Morice, 1582

The arrival of the **water-mill** in Europe around the 7th century (see p. 95) undoubtedly made civil engineers think of using **tidal energy**, for example by watching the **movement of the tides** in some narrow channels. In the 11th century there were so-called **tidal mills**, at least on the north coasts of the Adriatic and on the English coast. They then multiplied along numerous European coasts, despite the problems posed by the **variable heights of the tides**. In 1582 the German engineer Peter Morice (or Moritz) installed a tidal mill near London which activated a **force pump**; this must have been powerful, for it provided London's water. It was only at the end of the 19th century that the idea of tidal power stations again held the attention of the engineers, notably because of the great works which had been undertaken to master the problem of flooding in certain estuaries. At the end of the 1980s the idea of tidal power stations was thought to be largely underexploited.

Turbines (hydraulic, gas and steam)

Hero of Alexandria, 1st century BC; Leonardo da Vinci, c.1480; Ramelli, 1588; Branca, 1629; R. and P. Sualem, 1682; Segner, 1750; Euler, 1754; Barber, 1791; Burdin, 1824; Fourneyron, 1827; Avery, 1831; Poncelet, Howd, 1838; Thomson, 1840; Stolz, 1872; De Laval, 1883; Parsons, 1884; Rateau, Curtis, 1894

The principle of the turbine, a mechanism which collects the kinetic energy of a fluid, transmits it by means of a shaft and transforms it into another form of energy, is closely linked in origin to those of the **steam engine** and the **mill**; the sources they have in common are in fact Hero of Alexandria's **aeolipile** (see p. 105) which was made in the 1st century BC and used a steam jet to turn small blades fixed to a shaft, and the mundane **water-mill**. Indeed their derivatives sometimes overlap, such as the **hydraulic piston mill** which was invented by the Chinese in the 6th century and the description of which would later be carried to Europe by travellers to inspire the first attempts at the **steam piston engine**. For centuries, however, the principle of the steam engine remained dormant and that of the mill was used. At some stage, however, probably around the 16th century, this common stem was to split into two divergent branches. That of the steam engine was to yield only more advanced steam engines, with a later shoot linking the principle of the cylinder and the piston to result in the **four-stroke engine**. That of the turbine was to provide machines which functioned, depending on the type of model, on hydraulic, gas or steam energy. In the 20th century a sudden spread of gas turbines would lead to the **turbo-compressor**.

The great technological hiatus which followed the end of the Roman Empire and the Graeco-Roman culture hardly benefited the development of Hero's remarkable invention. The first to take up the principle again theoretically, but in a different form, was Leonardo da Vinci in 1480. He drew a sketch of a propeller fixed to a vertical axle which was turned by a flow of air rising from a chimney; the axle activated a transmission gearing system which itself operated a pulley, the belt of which could be used to turn a roasting spit. This first attempt at a **hot-air turbine**, that is, a gas-powered one was called the **smokejack**. In 1629 the Italian Giovanni Branca described an assembly which today seems extravagant and in which a steam turbine brought an impossible series of gears into operation to activate a rolling mill. Modern mechanical knowledge has revealed that the friction of these gears would have absorbed the small amount of energy generated by the steam. Perhaps Branca's description lacked precision, for the **rolling mills** certainly operated on this principle in Germany and Spain at the end of the 17th century. The idea survived nevertheless, for in 1603 another Italian, Vittorio Zonca, proposed to turn a roasting spit — again — using a device which was similar to those of both da Vinci and Branca. One cannot help thinking today that this steam mill was a very persistent idea, for nearly a century later it was taken up again, or possibly reinvented, by the Englishman John Wilkins (for technological projects seldom spread). Later still it reached an odd stage which was described in a rather obscure way in the 17th century by another Englishman, John Dumbell. Dumbell seems to have had the

> It is worth noting that nearly twenty centuries passed between the prototypes of all the turbines, the aeolipile and their first industrial applications. Notwithstanding the successive states of technology, the steam turbine in particular could have existed well before the time it did; engineers became interested in it only when energy requirements became urgent.

idea of a gas turbine with a shaft of several rows of blades which would work by the burning of alcoholic vapour; this would therefore have been a gas turbine prefiguring the **modern combustion turbines**. It was the Englishman John Barber, however, who in 1791 came up with a rather more rational project for the gas turbine. It consisted this time of an obviously very basic device which produced hot air by burning fuel; this air was compressed and driven out by a turbine wheel with blades. Barber had actually done nothing more than reinvent the aeolipile, (see p. 105) and according to modern reconstructions of the invention, it seems to have had an almost non-existent or mediocre output; nevertheless, it did have the significant advantage of including a **compression chamber**.

During the following decades several derivatives arose, but it was only in 1872 that the German F. Stolz eventually made some progress. He proposed a turbine, also with a combustion chamber, from which the hot air was directed towards a **heat exchanger** where it would be reheated again using air coming from another combustion chamber. It would then be directed towards a compressor activating a paddle-wheel which would send it out to the open air.

Stolz had in fact just devised the principle of the **double cycle open gas turbine**, but the technology of the time did not allow him to put it into practice. This turbine or a similar model was actually not to be made until 1903, by the Frenchmen R. Armegaud and C. Lemale. The gas turbine had no successful conclusion until the end of the 19th century.

The **hydraulic turbine** was to be more successful, for the simple reason that it was directly derived from the water-mill, of which there had been good experience for several centuries. While they still had trouble in mastering the **mechanics of gases** and the mysteries of steam, and were even less successful with the technology that they implied, the engineers of the Renaissance and the subsequent centuries were much more at ease with problems concerning water. Hydraulic pumps had been around since ancient times (see p. 000) and the hydraulic motor, which was introduced in the 16th century by Ramelli, had familiarized them with hydraulic energy and ways to exploit it.

The first prototype of the hydraulic turbine which really worked was manufactured on a large scale; this was the enormous set of 14 water-wheels built in 1682 by Rennequin and Paul Sualem for supplying water for Marly and Versailles. Being fed by water from the Seine the 'Marly machine' activated no less than 259 pumps. Without a doubt it was not simply a collection of mills but a true turbine, for it really did transform hydraulic energy into mechanical energy in the pumps. In 1750 the German Johann Andreas von Segner came nearer to the modern conception of a turbine by designing a **reaction turbine** which consisted of a vertical cylinder with radiating horizontal blades; this was very similar to the shaft of Dumbell's gas turbine. Four years later the famous Swiss mathematician and physicist Leonhard Euler published a theory of turbines and designed a model which would raise the output of the turbine — the word had not yet been invented — by changing the flow of water to be against the profile of the wheel; in order to do this the water was distributed using a cogwheel with a conical section inside a fixed cylindrical ring.

This was the idea that the Frenchman Claude Burdin adopted and modified in 1824. Here the distributor was inside the wheel. It was his pupil Benoît Fourneyron who made Burdin's turbine in 1827 and also coined the term 'turbine'; the experimental model generated 6 hp under a 1.4 m fall of water. Fourneyron, for his part, had invented **variable angle paddles** which optimized the output by adapting to the height of the falling water. The first Fourneyron turbines had a rotor measuring only 31 cm in diameter and turned at 2300 rpm. Hydraulic turbines therefore became a reality. They were improved considerably as a result of the following principle: instead of putting the paddles on the sides of the wheel, they were placed as close to the centre as possible. This type of centripetal turbine was designed by the Frenchman Jean Victor Poncelet in 1838, made in the same year by the American Samuel B. Howd, and then improved in 1840 by the Englishman James Thomson who invented the adjustable distributor. Later modifications of the turbine were mainly to do with the shape of the blades and the distribution system, in order that they could be equally well used in both low and high waterfalls. Indeed, hydraulic turbines allowed the **production of cheap electricity** — hydroelectric power — as attested by the first three-phased transportation of electricity in 1891 over a distance of 175 km. This demonstration took place at the International Exhibition of electricity at Frankfurt-am-Main in Germany. Thus by the end of the 19th century the turbine was expanding fast.

The **steam turbine**, the most direct descendant of the aeolipile, did not appear until very late in technology, at the time when hydraulic turbines had already reached maturity. The steam engine was already occupying the imaginations of many, and when the American William Avery made a weak steam turbine in 1831 it was virtually no more than a reconstitution of the aeolipile, apart from having two exhaust pipes instead of four. Fifty of them were made for use in sawmills and worksites, but as they were noisy, fragile and difficult to adjust, and no one was interested in improving them, they were abandoned for a time. It was the Englishman Charles Parsons at the beginning of the 1880s who brought this type of machine to life by having the idea of letting the steam out in stages, which made it work more smoothly. In 1883 the Swede Carl G.P. De Laval modified the steam turbine in the following way: he directed the steam jets through four fixed, convergent openings towards a rotating disc with blades on it. This turned in the direction of the jets but dispersed the steam in a divergent manner; in this way De Laval made turbines which turned at 42 000 rpm and which he developed as maritime motors. Meanwhile in 1884 Parsons had made a turbine which had a central input of steam and divergent ejection, which enabled him to make reliable motors. In 1894 the Frenchman C.E.A. Rateau and the American Charles G. Curtis independently improved the output of the De Laval turbine by introducing the principle of **pressure stages** in the supply of steam. Curtis invented the **compound turbine** which had two concentric wheels: the steam ejected by the first provided the input for the second. Thus at the end of the 19th century the steam turbine was a success too.

Universal joint

Fang Feng, c.140 BC; Cardano, 1550

The mechanism enabling the relative angular displacement of two shafts of which the geometrical axes converge at the same point, in order that movement in all directions can be made, is the basis of the modern **gyroscope** and of a number of aerial navigation systems which enable the definition of the angle of an aeroplane in relation to the horizontal. Also known as the **cardan**, its name is derived from that of the Italian scientist Girolamo Cardano who described it in his 1550 treatise *De subtilitate*; for a long time the invention has been attributed to him. The **cardan** is indirectly mentioned however in a seduction scene of the *Ode to Pretty Ladies* by the Chinese Ssuma Hsiang-ju (Sima Xiangru), in the form of a perfume burner, the model of which, being very old, is known: it consisted of a set of interlocking circles which were effectively linked to one another by pivots with converging geometric axes. It was invented by a man named Fang Feng in the 2nd century BC. The cardan is mentioned again in more detail in another Chinese text, dating from 189. This spoke of Fang Feng's secret being lost, but it was found again by a certain

Ting Huan (Ding Huan), to whom the reinvention of the cardan is attributed, may also have been the inventor of the magic lantern (see p. 44).

Ting Huan (Ding Huan); at this time the cardan was used only in the making of perfume burners, for it meant that they could be hung from the ceiling without being under any risk from the breeze. There is good reason to think that the braziers which were given to the empress Wu Hou (Wu Hu) in 692 were designed on the same principle for they could be pushed in any direction without the burning coals tipping out. It was the astronomer and mathematician Robert Hooke who modified it in the 17th century to make a **broken joint**, also known as a **universal joint**.

Watch winder

Beaumarchais, 1755; Bréguet, c.1780

Before the winder was invented the spring in a clock (or watch) had to be wound with a special key after the clock had been opened. In 1755 the playwright Antoine Caron de Beaumarchais, who was also an inventor and a skilful clockmaker, designed a watch for the Marquis of Pompadour which could be rewound without being opened by turning a ring mounted on the dial. The invention was renewed or remade by the famous Swiss clockmaker who was employed at Versailles, Abraham Louis Bréguet. He placed a knob on the edge of portable watches which enabled them also to be wound up without being opened. It was Bréguet too who invented the **universal winder**, a key which could be used to wind all timepieces and clocks from the outside. In 1842 his compatriot A. Philippe invented a **stirrup winder** which could change the hands when it was placed in a certain position.

Industry, Construction Materials, Civil Engineering and Tools

The technology which dominated invention in the domains of industry, materials, tools and civil engineering from ancient times to about 1850 was incontestably metallurgy. Consisting of a host of discoveries and actual inventions in which the two are often difficult to tell apart, it led to the manufacture of tools which controlled the development of civilization, beginning with the cast-iron ploughshares. Thereafter the ground could be worked to greater depths and better harvests obtained. Next, tools made of cast-iron and then of steel allowed hard materials such as stone to be shaped. Many great structures were built, in which the Romans displayed incomparable genius.

Glass, which probably resulted from a discovery, benefited from a series of inventions. Some of these were early, such as glass-blowing which dates back to the 1st century, and some were later, such as the hard lead oxide glasses. Glass was to play a primordial role in the realization of instruments for observation and measurement, which are essential for learning about the world. After metal, it was a key material on the road towards civilization.

Weaving, a major industrial technique, was not to be developed until steam and then electricity had been mastered. One of its most unlikely derivatives is computing, which was based on the perforated cards which Vaucanson tried to adapt for the weaving of complex designs.

One major revolution which is still not understood in the 20th century is that of the Gothic style of architecture. After being analyzed at length solely from a stylistic point of view, it introduced the integrally new concept of the supporting structure, which heralded modern architecture. Walls became thinner and more light could enter the buildings: the revolution of the incorrectly named 'Gothic' style also hinted at an imminent major cultural revolution.

Without doubt the reader will meet with a few surprises, particularly regarding the great ages of concrete, aluminium and drilling towers.

Adhesives

Anon, Egypt, 3300 BC

The development of adhesives constitutes one of the best examples of a continuous invention. The first evidence of adhesives is given by an Egyptian bas-relief dating from 3300 BC which shows some Egyptian cabinetmakers applying bands of wooden veneer to a bracket, probably of sycamore.

It was not until the second half of the 20th century that the mode of action of adhesives was understood; it is in fact due to numerous factors, of which **molecular attraction** is at the top of the list.

The oldest adhesives were based on **plaster**, **resin** (often blended), **starch, fish glue, casein, gelatine, wax, egg-white** and **bitumen**. The Egyptians used starch for sticking papyrus fibres. Numerous primitive civilizations, such as those of Oceania, also used **blood** (the adhesive action of which is due to the albumin), for example for sticking the skins of drums on to the cylinders. The recipes have varied throughout the years according to the different civilizations, trade associations and uses. **Synthetic glues** were not to appear until the 1930s.

Aluminium (extraction of)

Œrsted, 1825; Wöhler, 1845; Sainte-Claire Deville, 1854; Hall, Hèroult, 1886; Bayer, 1890

Aluminium has been known since antiquity and perhaps even before then, for after oxygen and silicon it is the most widespread element in the Earth's crust. It is found particularly in the kaolins with which the people of Mesopotamia made fine pottery about 7000 years ago, and twenty centuries before Christ the Egyptians and Babylonians tended to incorporate **aluminates** into various products in their respective pharmacopoeias. Aluminates are, among other things, gastric antacids (and potential poisons for the nervous system). Although its industrial ore, **bauxite**, is relatively rare, aluminium is frequently found in the unrefined form of salts, such as the aluminium or alum sulphates which the Romans habitually used from the 5th century BC for the fixing of cloth dyes. This use explains the origin of the modern word, for the Romans referred to the aluminium salts as *alumen*. It is not completely out of the question that the metal might have been extracted from it in the past, but it was only in the 18th century that it was eventually realized that aluminates contained a metal.

There is some reason to believe that the Romans knew aluminium as a metal and that they even used it for making objects. In Chapter 26 of Book XXXVI, the penultimate book of his *Natural History*, Pliny the Elder recounted an anecdote which was to be repeated many years later by Isidore of Seville in Book XVI of his *Etymologies*: in about the first year AD a Roman craftsman presented a shining, lightweight cup made out of a new metal to the emperor Tiberius. Some writers see this as proof that aluminium had already been discovered; it is not unlikely, but it does not categorically exclude the possibility that the cup in question might have been made of a particular alloy, for example tin and antimony.

In 1807 the famous English chemist Sir Humphry Davy applied himself to the extraction of aluminium and although he did not succeed he became convinced that a new metallic base did exist there. This he called *alumium* at first, and then *aluminum*.

The first to succeed in extracting it was the famous Danish physicist and chemist Hans Christan Œrsted or Ørsted; in 1825 he noticed that the amalgams obtained were similar to tin in sheen and colour. Twenty years later the German Friedrich Wöhler managed to bring about the extraction again, and was also the first to establish the characteristics of the metal in terms of specific weight, ductility, colour, etc. All the same, the aluminium was still only a curiosity and its extraction was a very expensive process. Taking up Wöhler's technique, the Frenchman Henri Sainte-Claire Deville managed to melt the fragments, which were the size of pinheads, into balls and then into ingots. The first aluminium ingot was exhibited the following year, in 1855, at the Universal Exhibition in Paris. This time the industrial advantages of the new metal were clearer. However its actual destiny only began when the American Charles M. Hall and the Frenchman Paul Louis Toussaint Héroult independently carried out the extraction of aluminium from bauxite by electrolysis. If they were able to improve the Sainte-Claire Deville process, it was due to the increasing availability and progressively decreasing cost of electricity. It was Héroult in 1888 who was the first to make the bronze-aluminium alloy and thus invent **duralumin®**. Two years later the German Karl Joseph Bayer completed a method of extracting pure alumina from ores which had a low bauxite content.

Arch

Anon, Egypt, 3000 BC

The invention of the arch is fundamentally important in architecture. Indeed, it laid the foundations for the monumental architecture which the Romans were the first to carry to great heights. Traditional teaching holds that the arch and the **vault** were invented by the Romans, which is not the case at all. The oldest example of the arch dates back to 3000 BC and is found in a tomb which was uncovered in Hèlouan in Egypt (near Cairo). The arch was made using flat uncooked clay bricks which rested on the walls, with the rows of bricks being arranged radially and the circular arch interstices being filled in with mortar strengthened with pieces of pottery. An identical technique was used at the same time in Mesopotamia and then extended, some 500 years later, for the building of arched porches. To increase stability these were sometimes built using two layers of superimposed bricks which slanted in opposite directions. At Tell el Rimah, also in Mesopotamia, the span of the vaults in a temple dating from the first half of the second millennium BC reached 3.8 m, which is proof of the expertise of the architects in this domain. Two elements worked towards the stability of these vaults. The first was their relative lightness, due to the fact that more chopped straw was incorporated into them than in ordinary bricks; the second was that grooves were made in them while they were still soft to facilitate pouring in the mortar (liquid mud) between the adjacent surfaces.

The first Egyptian vault was of the type called radial, but it was built with parallelpipedic bricks. The next major stage was reached in 675 BC when the architects of Tell Jemmeh, which is now in Palestine (they were Syrian architects as the country was under occupation), built the first **semi-circular vaults** using bricks which were **curved sections of a circle** (the biggest being 30 cm of the arch, the smallest, 24.5 cm), or **voussoirs**. It is impossible to determine whether this innovation ensued from a technique which began in Persia which, with astonishing boldness, used only two rows of voussoirs each with a range of half a vault. This technique is all the more surprising from the fact that bricks made of uncooked clay and measuring 120 cm in length were used. The impossibility of establishing a chronological relation stems from the fact that the Persian building where this technique appeared cannot be dated exactly (between 750 and 600 BC), whereas the Tell Jemmeh is precisely dated. However, examples of voussoirs are rare in pre-Roman architecture in the Near and the Middle East. This rarity remains somewhat strange, for the proof of their architectonic superiority ought to have caused them to become widespread.

Blade

Anon, Africa, 2 600 000 years BC

The blade is the oldest of all instruments; the oldest ones known were discovered in 1969 in Africa and date back 2 600 000 years BC. They were shaped from flint and had multiple uses: cutting, scraping, chopping and digging, as well as the usual function of slicing. These first blades gave rise to other types, the main ones being **side arms**, **axes**, **adzes** and **billhooks**, and finally **scissors**.

Brick

Anon, Sumer, 3rd millennium BC; Greece, 4th century BC

The brick seems to have been invented in Sumer during the 3rd millennium BC. It was an **unfired brick** which was left to harden in the sun and with which thick walls were built, using clay or bitumen as a mortar. This material apparently resisted the rain better than it did the action of the sun, which resulted in it becoming crazed. The bricks were prepared in wooden moulds. This was the only type of brick known in antiquity until the Greeks invented the **fired brick** in the 4th century BC.

The use of the brick, unfired or fired, depended firstly on what other materials were available in the area in which the building was taking place. In Sumer, as in the whole of Mesopotamia, there was no wood or stone, and so the brick was widely used, since it constituted the principal building material. In Egypt, as well as in America, where stone was available, the brick was only a secondary material, for use in the people's dwelling houses. The Greeks themselves did not make much use of the fired brick. The increase in population and the rising cost of stone nevertheless assured the growing success of the fired brick, which even found its place in certain parts of noble buildings, such as the cathedrals at Lübeck and Ratzburg in the Middle Ages.

Bridge

Anon, Mesopotamia, 4000 BC

It is very difficult to assign a date to the construction of the first bridges; among other hypotheses, it is possible that they were first made in the Stone Age by the inhabitants of **lakeside settlements**. This would have meant following the same principle as that used in the building of their houses: **piles** were driven into the mud, stabilized by ropes, and topped with **gangways**. Then, at another uncertain date, the **cantilever** bridge appeared, the oldest specimens of which were built in India. It had pillars built on each side with planks ballasted with stones placed on top of them; the successive rows of planks were adjusted to project outwards until the two parts met in the middle. The first signs of civil engineering in this domain seem to have appeared in Mesopotamia around 4000 BC. With bricks having succeeded the planks by then, following the principle of **corbelled construction in mortar**, the Mesopotamian engineers were the first to devise the **radial arch**, where the bricks were arranged in a radial fashion. This technique evolved during four millennia in Egypt, Babylon, Greece and Persia, resulting in the magnificent stone bridge at Martorell in Spain (219 BC), and the eight bridges, also in stone, which spanned the Tiber, and which were built between 200 and 219 BC. The Roman civil engineers exploited the principle of the radial arch admirably, and in the year 14 they made their largest aqueduct, the Gard bridge. This followed the principle of semicircular arches (6 on the first level, 11 on the second and 35 on the third). They used other techniques equally successfully, for example in Trajan's building of the Danube bridge in 104. This followed the principle of semicircular wooden arches which were supported by stone pillars (a principle weakly linked to that of the arch); the reach of these arches (of which there were 19) could be 51 m, an amazing achievement for the time.

The first ever metal bridge was the one built over the Severn in 1779 by the Englishmen Abraham Darby and John Wilkinson; it had a span of 33 m. The advent of industrial iron- and steelmaking meant that the ancient principle of the cantilever could be taken up again. The first person to exploit this was the German Heinrich Gerber in the bridge crossing the Main at Hassfurt; its average span was 129 m.

Concrete

Anon, Rome, 1st century BC

It was at least from the 1st century onwards that the Romans used concrete in their public works. The biggest venture to be identified to this day was the construction of the port of Caesarea by Herod the Great, where the engineers poured 50 t blocks in the open sea, inside wooden cases. The concrete, which was made of Italian volcanic ash, set in the high seas.

Crystal

Anon, Venice, around 1450; anon, Lyons, 1508; Ravenscroft, 1674

The first articles made of glass were dull, tinted and had poor **transparency**. In the 3rd millennium BC, the Egyptians used **rock crystal** for certain objects to which they wanted to lend a particular transparency, such as the eyes of the *Crouching scribe* (Louvre). This material never ceased to arouse the emulation of glassworkers throughout the succeeding centuries. In the Middle Ages, with cutting techniques being much more sophisticated, blocks of rock crystal were hollowed out to make precious articles, goblets, water-jugs, etc. From the beginning of the 15th century the Venetians succeeded in making a very white glass which had exceptional transparency for the time. It was made from a strict selection of

In **crystallography**, the glass known under the commercial name of crystal is not actually a proper crystal, but an **amorphous solid**.

base materials, **silicon**, **potash** and **soda**, particularly soda, which they imported from Egypt and Syria. The mastery of this type of glass, called **crystalline glass**, was attained around 1450, assuring the Republic of Venice a dominant position in this field. The secret behind it was discovered and its manufacture began in Lyons in 1508. At the end of the 16th century, still seeking a transparency and brilliance comparable to that of rock crystal, the glassworkers from Bohemia replaced the soda with **potash**, thus obtaining a better **brilliance**.

In 1674 the Englishman George Ravenscroft decided to incorporate some powdered **black flint**, similar to rock crystal, into the glass, and this enabled him to attain a **brilliance** and a **refractive index** comparable to those of the mineral. These fragments tarnished after a few months, however, and so Ravenscroft added **lead oxide**. This substance had long been known to make glass easy to work and therefore provide a more uniform end product. When it was added to the black flint it then produced a glass with a better **refractive index** than any other known glass.

Distaff

Neolithic

The first documents to feature the distaff, a wooden spindle on to which fibres are wound after they have been disentangled and twisted between the fingers, came from dynastic Egypt about 3000 years before Christ, but it probably actually dates from the Neolithic period (from 5000 BC). The first threads ever used seem to have been made of wool and they orginated in Mesopotamia.

Drill

Anon, Europe, before 1425

The drill is derived from **awls** for piercing holes, which were first made out of stone and bone, and then metal. Nowadays the name of **drill** is given to a bit-brace which has a point at one end and a free handle at the other, and which is turned by hand. It first appeared on a European painting dated 1425 which depicted the carpenter Joseph, the husband of the Virgin Mary, in his workshop. The first drills were made of wood but in 1505 in Germany drills made entirely of metal were used.

Drilling tower

Anon, China, 1st century

The first drilling towers in the history of technology appeared in China in the 1st century; they were designed not for the extraction of oil, but for that of salt taken from underground brine. This came up under the natural pressure of water, for the wells drilled by the Chinese were **artesian wells**. The drilling towers were up to 60 m high and were evidently built of bamboo. The equipment was simple: it consisted of bamboo cables at the end of which were cast-iron **drilling heads**. They were powered by human strength, with the drilling head being attached to a lever which was lifted by teams of workers. The head was raised to a height measured by the foreman which depended on the progress of the drilling, and fell back down upon the rock which it was to break. The operation could get underway after a small hole of a few tens of centimetres in diameter had been made using a spade. The drilling progressed at the rate of 3 to 90 cm each day, depending on the nature of the rock; the deepest drilling in Chinese history achieved by this process measured about 1350 m. As the drilling took place, **watertight joints** were driven down into the well in order both to stop it caving-in and to prevent unwanted entry of water. These joints were made of very large bamboos. The advantage of watertightness lay in the possibility of collecting **natural gas** wherever a pocket of it was encountered by the drilling head. Equally amazing was the process used for the **clearing** of the well: it consisted of **suction** through a leather sheath by means of a **two-cylinder pump** such as the sort used for the flame-throwers. There were a great many of these drilling towers. In 1089 they were limited by imperial edict to 160 in the prefecture of the province of Ch'eng-tu (Chengdu) alone. Such a restriction can be explained by the fact that since the salt trade was so lucrative it was often carried out in secret, although the salt industry had been nationalized since the 2nd century. Drilling towers in China continued virtually unchanged until the end of the 19th century.

The 1828 account by the French missionary Imbert on Chinese drilling towers left the West all the more incredulous that the European experiments, which were carried out using chains of rivets, had been disappointing. The first application of the Chinese technique in the West took place in 1859 at Oil Creek in Pennsylvania, and was the work of the American E.L. Drake. It was not derived from Imbert's account, but from the knowledge of the Chinese travellers who had been introduced to the United States to be employed doing extensive construction work, particularly on the railways. This technique was not modified until the advent of steam and the **rotary head**.

Fibreglass

Riva, 1713; Libbey, 1893

In 1713 the scientist René de Réaumur presented a fabric made entirely of fibreglass to the Academy of Science in Paris. It was the first known example and so had probably been made by his manufacturer, the Venetian Carlo Riva. The invention must not have been followed up, for it was not until 1893 that the American Edward Drummond Libbey succeeded in spinning glass fibres; being woven with silk, the material produced was extraordinarily successful. It was at the beginning of the 20th century that interest arose in the industrial uses of fibreglass and the manufacture of it was improved.

Libbey made a gift of a dress of glass fibre and silk to the child Eulalia of Spain. It was no more than an elegant curiosity, for the material was stiff and tended to break and so it could not withstand being worn, let alone folded.

File

Anon, Egypt, 1500 BC

The first known file was made of bronze and was found in Egypt; it dates from 1500 BC. The Greeks used **shark skin** for burnishing. The file does not seem to have appeared in Europe before the 4th century BC.

At the beginning of the 16th century, Leonardo da Vinci designed a machine for making files, but the idea was not applied until 1758.

At the end of the 16th century, files were used in the making of screw threads, because at that time screws were made by hand.

Gothic (style)

Anon, France, c.1120

The Gothic style is both one of the major inventions in the history of architecture and the incontestable source of modern architecture. It supplied a **support structure** other than the **supporting wall**, the reliance on which had imposed such architectronic constraints. This was made possible by the equal distribution of the pressure exerted by the vaults on the segments of the opposing broken arches which were supported by pillars (**ribbed vault**). From then onwards the supporting function of the walls diminished and the buildings could rise to great heights and have enormous openings — stained glass windows. The Gothic style also included the use — and the invention — of **abutting arches**, which were meant to respond horizontally to the deviating vertical pressures exerted on the pillars.

The technological revolution of the style known as Gothic (and which owes nothing to the Gothic culture) did not take place all at once, but by stages. Nevertheless, it took place at a surprising speed, given the relative slowness of technical progress at the time. Indeed, its premises appeared in Lessay in 1100 and its first major realization, the Cathedral of Saint Denis, dates from 1140. It was in fact at Lessay that the principle of **diagonally crossing supporting arches** (**groined vaults**) began;

Several countries in past years claimed the authorship of the Gothic style, until the Englishman G.D. Whittington proved in 1809 that the style had been invented in Ile-de-France. It has been suggested that the idea of the diagonal rib was brought back from the Orient because of the crusaders and that it proceeded from the **lancet arch** in Syrian architecture; whatever the case, the Syrian lancet arch never gave rise to the pointed arch. The term 'Gothic' is derived from the expression *alla tedesca* which was used pejoratively by the Vatican to define the new style which was judged to be 'barbaric', hence 'Gothic'. In fact the disapprobation directed towards the Gothic style by the papal authorities seems to have been inspired by dematerialization of the buildings, because of the slender pillars and the vast stained glass windows; for the Roman theologians, this revolution verged dangerously on mysticism and the refutation of the Incarnation, and they went as far as saying that the Gothic edifices had been built by freemasons, Jews and pagans, and such an undertaking would never have taken place in Rome.

these were still semicircular arches but in a Romanesque architecture they already had a supportive function which heralded the **pointed vault**. Since they were not very satisfying aesthetically, because they rested lower down than the transverse and longitudinal arches, ten years later at Saint-Etienne of Caen they evolved towards their characteristic supporting function in a **six-part crossed rib vault**.

In 1155 the characteristic broken arch or **ogee** appeared in the porch of Saint Peter de Moissac. Five years later the first **transverse ogival arches** appeared at Saint-Etienne de Beauvais and at Morienval. These were properly ribbed, in contrast to their predecessors, and they appeared respectively in the transept and in the cloister of the above churches. The English had already attempted the use of ribs in Durham in 1093, but it was in France at Saint Denis that the astonishingly complex calculations of resistance in the Gothic style were eventually mastered. Several 'repetitions' on a lesser scale undoubtedly contributed to this masterpiece, such as the choir of Saint

***The tower of Notre-Dame Cathedral** in Laon. Drawing in* Album of Sketches *by Villard de Honnecourt, first half of 13th century*

Germer, near Beauvais (1132) and, previously, the chapter at Jumièges (1109). Thenceforth followed the choir in the cathedral at Sens, which was in six parts like the ones at Saint Etienne de Beauvais, Noyon (1150), Laon (1160), Notre-Dame of Paris (1163), Chartres (1194), Bourges (1200), Rouen (1201), Reims (1210) and Amiens (1220), then the new naves in Saint Denis (1232), Sainte Chapelle (1243–1248), Troyes (1262) and Rouen (1318). Even though the ribbed vault had been used in Durham, then in Winchester in 1107 and in the crypt of Gloucester Cathedral (1120), the Gothic style did not make its first appearance in England in terms of a complete style, with the crossed rib vault, until 1174, when William of Sens rebuilt Canterbury Cathedral. The Gothic style dominated the countries in Central and Northern Europe where it supplanted Romanesque architecture, but it disappeared for a few centuries after the short-lived decadence of the flamboyant and decorated style, to be technologically discovered at the beginning of the 19th century when metallurgy again enabled supporting structures to be made, this time out of cast-iron (as in the Crystal Palace). Then in the first half of the 20th century architects used reinforced concrete to make supporting structures.

Lock (on a canal)

Ch'iao Wei-yo (Jiao Weiyo), 984

The use of a compartment with two gates, or **lock**, which enables a barge to pass from a lower to a higher level of a canal and vice versa, appeared in Europe in 1373. The invention of the lock dates back four centuries previous to this, however, for it was made in 984 by Ch'iao Wei-yo (Jiao Weiyo), an assistant transport inspector in the Hanan region. The invention and its realization have been widely described. The lock enabled the Chinese to make considerable extensions in the **canal network** across the Empire, for the engineers no longer had to take height differences into account.

Machine tool

Wilkinson, 1775; Nasmyth, 1851; Spencer, 1866 and 1895

It is difficult to attribute a definite date to the birth of the machine tool concept. If one understands this to mean a machine capable of carrying out a task without human intervention and by the harnessing of natural energy, this concept cannot be restricted to machines which use steam or electricity. In this case the **bellows** activated by a **hydraulic piston motor** which were invented in China in 530 constitute the ancestor of all machine tools, closely followed by the **hydraulic millstones**. However, the convention which joins the date of the birth of the machine tool to that of steam, that is, to the 18th century, is not arbitrary, for it was steam which introduced the idea of independent machines and which conferred to the tool the integrally new trait of mechanical regularity.

In this respect the first machine tool of modern times was the **reaming machine**, invented in 1775 by the Englishman John Wilkinson, which enabled the cylinders of Watt's steam engine to be made. It was not only to undergo a good many improvements, but also to serve as a prototype for numerous other machine tools, such as Eli Whitney's **milling machine** (1818), various **planing machines**, **copying machines**, **saws**, **filing machines**, other

milling machines, **grinding machines**, **rectifying machines** etc. Machine tools then began to be made for the manufacture of certain objects, such as the machine for the production of ship capstans, invented by Isambard Kingdom Brunel (1808). It is owing in particular to the strict criteria for the measurement of objects, which were imposed in 1834 by the Englishman Joseph Whitworth, that the machine tool could eventually contribute to the spread of **standardization** (see p. 137).

One of the key dates in this domain was that of the making of the first **steam power-hammer**, which was presented at the 1851 Great Exhibition in London by its inventor, the Scot James Nasmyth. The machine tool concept had not changed, but the unprecedented power and capacities of the instrument inaugurated the age of **heavy metallurgy**; thereafter objects of very large dimensions could be made, for bridge and tunnel structures, and for maritime and military engineering.

The beginnings of modern machine tools were provided by an **automatic lathe** invented in 1866 by the Englishman C.M. Spencer, which reduced the need for human intervention even more. In 1895 Spencer invented an even more sophisticated machine, the **multiple punch turret** which allowed several operations to be carried out at the same time.

> The machine tool is one of the inventions which has been instrumental in changing social structures. It has in fact enabled industrial production to increase while the number of hours worked has reduced, consequently lowering the **cost price of manufactured goods**. Moreover, it has meant a reduction in the number of workable hours, prefiguring both the age of leisure and the devaluation of manual work.

Metal construction

Bogardus, 1848; Saulnier, 1871–1872; Viollet-le-Duc, 1872; Nouguier, Koechlin, Lockroy, Poubelle, Eiffel, 1886

Eight centuries separated the revolution of **Gothic architecture** (see p. 123) from its modern corollary, constructions with metal frameworks. The reason for this was the often unjustified delays in the progress of **metallurgy**, as well as an excessive confidence of a cultural nature in the building techniques inherited from the Romans. **Cast-iron**, however, had existed long before the advent of metal construction in the last third of the 19th century and its advantages had been underestimated; it had been used by the 17th century classical architects, Gabriel and Perrault in **cramp-iron assembly** (which, moreover, had been practised in ancient times). However at the end of the 18th century cast-iron **posts** and **girders** appeared undoubtedly for the first time in architecture, in the flax factory in Shrewsbury in Great Britain, which was built in 1796–1797 by A.W. Skempton. More space was required for the new, bulky **power looms** and only iron enabled a large enough **span** to be obtained. Skempton and his successors, who adopted the method of cast-iron posts and beams, were not really inventors; rather they adapted a relatively new material to ancient techniques. They did not invent the **supporting structure** which is the distinctive feature of metal construction.

The next stage was reached by Henri Labrouste during the building of the Saint-Genevieve library (1843–1850) and then during that of the National Library (1855–1875). Despite his reputation as an innovator, Labrouste did not inaugurate a proper invention either. His was a heterogeneous system, firstly because in the Saint-Genevieve library the supporting structure was enclosed in a stone wall, and secondly because the central supporting columns were themselves supported by stonework pillars. The boldness of the formula of the Gothic Ile-de-France architects took a long time to come to light, therefore, although it was much easier to apply using metal which, for example, allowed the supporting elements of the building to be linked by tympanums. By the time Victor Baltard built Les Halles (1854–1857) and Jacques Isidore Hittorff built the Gare du Nord in Paris (1861–1864), both men finally making supporting structures which were almost entirely metal, the American James Bogardus had already erected a five-storey building with an entirely metal structure in New York in 1848. With the façade walls abolished and the time — and the genius — of the master Gothic glassworkers gone for ever, Bogardus settled for façades made of paned glass. An important point is that the building parts were **prefabricated**, and Bogardus and his successors could quickly put up similar buildings right across America: factories, warehouses, exhibition halls or railway stations. Bogardus therefore did the work of an inventor.

Almost immediately he was to have a follower, the Englishman Joseph Paxton, who built the famous Crystal Palace in London, one of the peaks of modern architecture. He put his previous experience of building huge glasshouses to good use in 1851 when he built an enormous building measuring 240 000 m², also using prefabricated parts.

Bogardus and Paxton, however, had only effected the juxtaposition of self-contained parts, which did not tie in with the spirit of architecture as a whole, and even less with that of Gothic architecture. It was up to the Frenchman Jules Saulnier to build the first **non-cellular metal structure building**, which had a real iron skeleton and not just cages joined together; this was the Meunier Chocolate factory in Noisiel. Its whole weight rested on four hollow square iron pillars; this development was to give rise to many levels of modern building. Saulnier's factory, built in 1871–1872, certainly owes a great deal to the theoretical views of Eugène Viollet-le-Duc, a great lover of Gothic architecture, who had been rudely maligned (because of the advanced restoration of Notre-Dame in Paris). He had been the first to understand the spirit of the Gothic architects and the way in which they used **thrusts** and **counter-thrusts**, being resolved in the invention of **intersecting ribs, ogees** and **abutting arches**. He was also the first to systematically use this lesson in metallic structures, inventing for the purpose the crucial concept of the 'organism'. This consisted essentially of two columns with each one holding back the thrust of the other by the interplay of **tie-beams, constraints** and **diagonal groins**, without which it would have been necessary to return to lead buttresses as had been done in the Hall of Industry in Paris in 1855. Viollet-le-Duc did not publish his theory until 1872, but it had been taking shape for a long time, and no credit is taken away from Saulnier and his brilliant application of it in the Meunier Chocolate factory.

All American architecture, beginning with the famous Chicago school, led by William Le Baron Jenney, took its inspiration for the first **skyscrapers** from Saulnier and Viollet-Le-Duc.

In the meantime, international attention was being caught by another event: the building of the Eiffel tower. This has often been presented in a totally erroneous way as the individual whim of Gustave Eiffel. In actual fact, for several years the imaginations of many had been contemplating projects to build a tower dominating a city. Trevithick, one of the 'fathers' of the **locomotive** (see p. 198) had already proposed to build a cast-iron tower 1000 feet high in the heart of England; the Frenchman Cabillet had planned another for Paris, and the Americans had already built the Washington obelisk (169 m). Two engineers from the Eiffel company, Emile Nouguier and Maurice Koechlin, presented a similar project to that of their patron at the 1889 Exhibition; Gustave Eiffel judged it to be imperfect. Nevertheless this project was nearly accepted by the Civil Service, and when he was requested to provide some ideas to the Trade minister Edouard Lockroy, Gustave Eiffel was caught unawares and so bought back Nouguier and Koechlin's idea. In 1886 Lockroy and the Prefect of Paris Eugène René Poubelle opened a competition for the construction of a tower measuring 125 m across the base and 300 m in height. Eiffel's project was chosen by the jury.

Metal pipes

Anon, c.3rd century BC

It seems to have been around the 3rd century before Christ that the Romans designed the pipe, for they were the great civil engineering entrepreneurs of antiquity as well as the greatest specialists in **water conveyance**. The pipe was originally made of baked clay and was used for supplying water in different areas of the towns of the Empire and into houses. In the 1st century the engineer Vitruvius codified the different ways of making the baked clay pipes, the thickness of which, he said, should be no less than 3.5 cm and the length of which varied between 1 m and 1.2 m. Vitruvius, without doubt drawing from his own long experience, also wrote down the recipe which ensured the watertightness of the joints (chalk mixed with olive oil), adding with characteristic Roman practicality that it was a crude but effective product. In the same text, however, Vitruvius also codified the manufacture of the first known **metal pipes**, which were made out of **lead**. Vitruvius was certainly not the inventor, for Hero of Alexandria also mentioned them in the 1st century BC, and lead pipes must have existed already, for without them Ctesibius would have been unable to make his organ. (see p. 70). As we have no proof of the existence of lead mines in Egypt, the merit of the invention codified by Vitruvius must be attributed to the Romans, who did have the use of an abundance of lead mines in the Empire. Vitruvius stipulated a thickness of 6.27 mm and a length of 2.95 m for the pipes, and they were made in ten different diameters.

Several specimens of pipes have been unearthed in diverse parts of the Empire and they show that the Romans made them by rolling bands of metal around a cylindrical or triangular mould, and then soldering the sections together.

Hero of Alexandria recommended pure **tin** for the **soldering**, but a **lead-tin alloy** was often used.

Vitruvius also defined the necessary conditions for preventing the two main dangers which threatened the **canalizations**: bursting due to pressure and blocking due to the accumulation of sediment. He recommended avoiding acute angles to prevent bursting, for these subjected the smaller pipes to excessive pressure, and he advocated enveloping the bends in red sandstone whether they were made of baked earth or lead. As for the deposits, these were usually eliminated by the use of **drainage tanks** into which the water was poured before being distributed; these tanks were long enough for the material in suspension to become separated by its own weight before it reached the openings of the adjoining pipes. These tanks were cleaned out regularly. Nevertheless, if these were calcium deposits, they must have formed inside the pipes, and Vitruvius's method for getting rid of them has not yet been fully understood. His term **colluviaria** may have designated an abrasive mantle fixed to the end of a flexible rod for use against the resistant sediments; however, the Romans are known also to have used a pneumatic pressure system similar to that of the pumps.

It was only in 1886 that the German Hannesman brothers invented a process for hot rolling of the pipes without soldering.

Vitruvius's writings on pipe manufacture provide a precious historical document. In them he actually stated a preference for baked clay pipes because, he said, lead pipes had the serious disadvantage of the formation of **lead oxide**, or ceruse, which is dangerous to the health. Several hypotheses have suggested that the fall of the Roman Empire could be explained by lead oxide poisoning, given that the Romans used lead piping and lead containers in winemaking. Vitruvius's mention indicates that the Romans were fully aware of the dangers of lead oxide, however, and it does seem that they consequently made their drinking water supply pipes in many of their houses out of baked clay, reserving the lead piping for other uses.

Metallurgy

Anon, Mesopotamia, 4000 BC; anon, Egypt, 2600 BC; anon, Hittites, Black Sea, 1200 BC; anon, China, 4th–3rd century BC; anon, India, 1st century; Bessemer and Kelly, 1854; Siemens, 1856 etc.

It is somewhat risky to use the term 'invention' concerning metallurgy, for everything points to the fact that the first known metals were discovered by chance in the native state, and the first metal objects were made simply by hammering bits of native metal. Besides, it is impossible within this framework to present all the inventions which have resulted in the birth of modern metallurgy; they number in the hundreds and most of them consist essentially of improvements to already existing inventions. We have therefore included those stages in metallurgy which are characterized by the creation of a new technique.

The first invention or 'semi-invention' in metallurgy seems to have consisted of the idea of the **hammering** of meteoric iron, especially nickel-iron, as well as copper and native ores. According to archaeological information acquired up to the 1950s, it took place in Mesopotamia, 4000 years before Christ. It was at the same time and in the same region that another fundamental invention seems to have taken place; this was the practice of heating the metal, in this case copper, to **melting point** and then **tempering** it by keeping it at a high temperature for several days in order to make it harder.

One of the most remarkable technological tricks in metallurgy was the invention of the gold-copper alloy **tombac**, which was carried out by the Indians from Colombia, or what is now Panama and Costa Rica. The alloy had a lower melting point than that of the two metals taken separately, and so it could be produced in low-powered furnaces; it used less gold and, by oxidizing the surface particles of copper, it meant that objects which had the appearance of pure gold could be made. The oxidation of the copper was effected by heating; then the blackened copper was cleaned in an acid bath of urine or citrus fruits and when it was polished, it gleamed.

Two other inventions of equal, if not greater, importance which also took place in Mesopotamia at the same time were to complement the basic picture of fundamental metallurgical techniques. One was the **reduction by heat or oxidation of ores** of copper or lead, or the method of extracting pure metals from ores by the process of burning or eliminating the impurities. The other was the preparation of **alloys**: it was towards the middle of the 4th century in fact that the Mesopotamians made copper-tin alloys, and this was not by accident, for some alloys had previously been made accidentally. These were the first intentional alloys in the history of metallurgy, and they inaugurated the Bronze age.

The extraction of a metal from an ore is worth a particular mention, for it was made possible by a specific invention, the **reduction oven**. This can be dated to around the year 3600 BC, but it is not known whether it appeared in Mesopotamia or Egypt. The reduction oven consisted of two superimposed chambers which were separated by a partition with holes in it. The ore was placed in the upper chamber, where the heat spread through the holes in the partition. As it slowly burned, the sulphur present in the ore was given off in the form of a gas, leaving only the metallic residue. This ore was impure, however, and according to deposits of it, contained iron, gold, silver, tin and lead. After being roasted the metal ore was brought to melting point and made into sheets which were hammered into shape.

Of course, the reduction oven was not used only for the extraction of copper. It gave rise to a secondary invention, that of permanent stone or iron **moulds** for the casting of objects to be manufactured in large numbers, or mass produced. In this respect, homage should be paid to the unknown genius who was the first to realize that copper is not found only in surface deposits in an unaggregated form but that it is also much more plentiful.

Four millennia before Christ, therefore, people had learnt how to use metal, how to melt and anneal it and how to extract it from the soil and combine it with other metals. Other techniques were to be added successively to this basic knowledge, which were sometimes inspired by the discovery of other metals. It was in this way that in the 2nd millennium, still in Mesopotamia, **cupellation** was invented as a result of the discovery of argentiferous lead. The process consisted of heating the ore in a **hearth furnace** to the moment that it lit up, when the silver became shiny. The lead, which has a much lower melting point, then formed a layer from the top of which the silver could be 'picked'. Before this, familiarization with the production of bronze would have also led the Egyptians to invent a casting process which is still used to this day, that of the **lost wax**. In order to make a metal figure, a wax figure was made first. It was then covered in clay and when the whole thing was heated, the clay hardened. Then molten bronze was poured into the clay mould to take the place of the wax which had duly melted during the baking of the mould. All that remained was for the mould to be broken.

The sixth and last invention during the 4th millennium was that of **solder** which was melted to join together different parts. The first to be made were gold-copper and lead-tin alloys.

Between the 3rd and 2nd millennia the techniques already known were extended to other metals. In 1300 BC a major event took place: iron began to be extracted from its ore. The first people to have done this were apparently the Hittites from south of the Black Sea; a century later the Philistines were to borrow the technique from them. Amazing though it may seem, at the same time this discovery inspired another invention, a **carbon-iron** alloy, that is, steel. It is not known whether the invention took place in Mesopotamia or Egypt, but it is almost certainly derived from observation.

The iron which was extracted from the ore came in the form of globules submerged in a semi-liquid magma; after being heated and subjected to successive hammerings, the globules — of nearly pure iron — were separated from the magma. All that remained was for it to be hammered. However a **soft iron** was then obtained which was of limited use. The first blacksmiths must have noticed that when the iron was put with wood it became much harder; this was in fact **carbon iron** and it had the additional advantage that it could be tempered.

At an unverified time, ten centuries before Christ according to some writers and only four according to others, the Chinese were the first to master the manufacture of cast-iron. This was a result of an invention which had an importance equal to that of the reduction oven: the **blast furnace**. Apart from their technological genius, the Chinese were well placed to design it: they had excellent fire clays for building the walls of the blast furnace as well as having a type of earth, which they called 'black earth', which was very rich in **iron phosphate** and which proved to be very useful in the manufacture of cast-iron: indeed, if up to 6% phosphorus is added to iron ore, its melting point is lowered from 1130 °C to 950 °C. This meant that they could attain the melting point of iron three to four centuries before the appearance of the first blast furnaces. When they were equipped with blast furnaces they no longer had any need for it. The Chinese also used enormous amounts of coal, and from the 4th century, perhaps even before then, they knew how to use it as a fuel. In order to make cast-iron they put the iron ore in crucibles in the shape of tubes and then put these in direct contact with the burning coal. The idea may seem to be disarmingly simple, but the fact is that the English had great problems with smelting of iron ore using coal right up to the 17th century.

The mastering of cast-iron stimulated Chinese metallurgy considerably and, as well as tools and weapons, they made the first cast-iron ploughshares which caused great changes in agriculture. In the 3rd century BC, one century after the manufacture of cast-iron, the Chinese succeeded in another masterstroke, the invention of **tempered cast-iron**: when it was kept at a very high temperature for a week, the cast-iron became much less breakable. This meant that the ploughshares could be improved again and that much larger objects could be made. Cast-iron thus began to resemble steel. In 1105 the Chinese even built a 26 m-high pagoda entirely out of tempered cast-iron by superimposing smelted octagonal storeys; it still stands in the Shandong province in Luoning.

In the 1st century the Hindus achieved another masterstroke in the invention of the process to manufacture **cast steel**. It was certainly steel, for according to the accepted definition, it contained no more than 1.7% carbon. In order to do this the Hindus hammered the iron to get rid of the slag, broke it up and smelted it with wood shavings which absorbed the carbon. The steel thus obtained was used for making tools and weapons. The same process was to be taken up by the Arabs in the Middle Ages to make the famous Damascene swords.

It seems to have been in Asia that the process of the **extraction of mercury by distillation** was invented around the 1st century. It is a strange fact that the Graeco-Latin world did not produce any invention in the domain of metallurgy. The first major invention to follow the Hindus' invention of cast steel was that of **puddling** by the Englishman Stephen Cort in 1784: this process consisted of decarbonizing the cast-iron in the blast furnace by agitating the molten mass. Consequently a mass of refined, pasty iron was obtained which could be rolled to make rods or plates, or poured into moulds.

In 1854 this invention was improved with the simultaneous inventions by the Englishman Henry Bessemer and the American William Kelly; these consisted of the direct purification of the molten iron by injecting air into a vat, or **converter**, the interior of which was coated with a refractory metal. By eliminating the surplus impurities, a metal was obtained which could withstand tempering and considerable stresses. Three successive inventions, the furnaces by the Germans Wilhelm and Friedrich Siemens (1856), the Frenchman Pierre Martin (1864) and the Englishman Sydney Thomas (1880), meant that all the iron ores could be used in the production of cast-iron which was then used in the manufacture of steel. The principle behind this consisted of preheating the air destined for the combustion of the impurities in the cast-iron, thus enabling the temperature of the treatment to be raised appreciably.

Mirror

Anon, Egypt, c.3000 BC

The mirror was in existence in ancient Egypt by at least around the 3rd millennium BC. The inventor is unknown but it is most likely that the object derived from the discovery of the reflective properties of a polished metal surface. The first mirrors

Numerous ancient authors such as Anthemius, Galen and Zonarus reported that Archimedes may have set fire to the Roman ships which were besieging Syracuse by focusing the Sun's rays on them using a hexagonal mirror. Zonarus wrote that Proclus repeated Archimedes's feat in 514 by setting fire to Vitalian's ships using bronze mirrors. These acts do seem rather doubtful as it would have taken a mirror of very large dimensions, and one which was articulated as well, to effect a sufficient concentration of light on the ships. The technology of the time does not seem to have made possible the manufacture of such mirrors. On the other hand, American experiments carried out in the 20th century have demonstrated that if the soldiers of Syracuse focused the light reflected from their polished buckles, which were probably made of bronze, they could have raised the temperature of the vessels to 300 °C. Ships besieging a town tended to come within some 30 or 40 m so that the targets were in range of their arrows. In any case, mirrors were not unknown to the Greeks, for Euclid had already published a theory on them in the 3rd century BC.

were in fact made of silver, gold, and bronze, and then tin and polished steel. The Romans and perhaps the Celts sometimes also used polished vitreous lava. The first mirrors were small and held in the hand, but in the 1st Century in Rome cheval glasses appeared which gave a full-size reflection of the body. It was in Rome too at the same time that the use of **polished glass** mirrors began to spread, behind which a sheet of metal, probably silver, was placed. This technique seems to have been lost and then rediscovered in the 13th century, in Germany or in Italy; it then gave rise to the type called **reflecting glasses** or **crystal mirrors**. It was in Nuremberg in the 15th century that **tin-plating** was discovered; this consisted of applying **molten tin** directly on to the glass. The technique was adopted in Venice which became a specialist centre for the manufacture of mirrors. Around 1675 the Norman glassworkers rediscovered the technique of **casting glass**, which allowed very large mirrors to be made. Twenty years later the Saint-Gobain factory acquired international renown by making these large mirrors. Until the 19th century mirrors were excessively expensive and the Hall of Mirrors at Versailles aroused as much admiration for the amount of money that must have been spent as for the splendour of its architecture.

Plane

Unknown origin and date

The plane and its longer forms, the **trying-plane** and the **jack-plane**, are rather mysterious instruments, in that their inventor is not known, which is quite often the case with tools, nor is the date they appeared, which is much more unusual. The only definite fact is that it cannot have been before the Bronze age because they are tools with a metal blade. The plane could possibly have derived from a similar instrument which would have had a flint blade and was used for neatening the surfaces of objects. The Egyptians do not seem to have used it around the 3rd millennium BC, for they rough-hewed objects using blocks of stone or wood with sand underneath; this was rough-hewing by **abrasion**. The Romans seem to have been the first to use the plane, and, paradoxically, like many other inventions, it disappeared for a long time after the fall of the Empire, to reappear only in the 17th century. The only fundamental modification it was to have was the invention of the **adjustment screw** for the blade, which was made around 1890.

Porcelain

Anon, China, 1st century

Porcelain differs from pottery in the fineness of its grain, its surface **vitrification** and its firing temperature (1280 °C instead of 500 to 1150 °C). It appeared in China in the 1st century, and was probably the fruit of research which had been pursued by craftsmen for eleven centuries. This had resulted in the progression from the already shiny **proto-porcelain** to proper porcelain by the selection of different clays and glazes, and the codification of the method of firing.

Power-hammer

Watt, 1784; Deverell, 1806; Bourdon, 1846

The invention of the power-hammer, a machine essential in metallurgy for the forging of large metal objects by force, is traditionally attributed to the Scot James Nasmyth in 1841. This was contested, however, no less traditionally, by the Frenchman Francois Cavé who had already used such a machine in his Parisian workshops in 1836.

In reality it was invented by the Scot James Watt, who patented it in 1784. Watt had imagined a hammer, the weight of which would be lifted by a **steam piston** and which would come down due to its own weight. It was too advanced an idea for the time, and had no more immediate success than that of the Englishman William Deverell, who also had the idea of accelerating the fall of the hammer by the injection of **compressed air** in the upper part of the cylinder (1806). For about thirty years before that the instruments used had been hammers with handles activated by camshafts, called '**tilt hammers**'.

The above-mentioned Cavé reinvented Watt's system, perhaps independently, and he patented it in his own name in 1836, being credited with having been the first to have the prototype of it built. Three years later Nasmyth, who had adopted Watt's idea too, still had not built a machine.

It owes its progress to the Frenchman Bourdon who in 1839 had a complete study made of the machine, which he was the first to call the power-hammer; it was built — not without reservations — by the Schneider brothers in their workshops in Creusot in 1841. The first power-hammer with its **falling mass** of 2.5 t from a height of 2 m was used in the manufacture of large forged parts. Bourdon rapidly extended the use of it in 1846 to **stamping** (see p. 136).

When visiting the Creusot factories in 1842, Nasmyth was all the more struck by the realization that Bourdon and Eugène Schneider had visited his workshop in 1839 and taken information from his sketches. Nasmyth took steps to have his rights to the invention recognized and an argument ensued. However the Schneider brothers explained to Nasmyth that the structure of his invention was too weak to withstand the impacts produced, a shortcoming which Bourdon had rectified. This was in fact an additional example of an invention which was 'in the air' and the true originator of which was an earlier worker, in this case Watt, or Deverell.

Screw

Archimedes, 3rd century BC; anon, Europe, 15th century

Archimedes can be credited with the invention of the **screw thread** [in the 3rd century], which led to that of the **endless screw**. This was widely used by the mechanics of the Alexandrian School and subsequently commonly in the manufacture of **presses** and **wine-presses**, but the invention of the screw itself cannot be attributed to him. The first screws appeared in the 15th century, connected with **nuts**. Their inventor is not known. These screws had hexagonal or octagonal heads without grooves and were tightened using keys; they have been seen in some 15th- and 16th-century armour as well as in some 17th-century clocks. The **wood screw** did not appear until around 1550, and, paradoxically, the screwdriver itself did not appear until two centuries later, in 1744 to be precise; its inventor is unknown too. Nor is it known exactly when the making of the thread by filing was superseded by the manufacture of the screw in the foundry.

Sewing machine

Wiesenthal, 1755; Thimonnier, 1830; Hunt, 1832–1834; Howe, 1844; Wilson, 1849; Grover, 1851; Gibbs, 1856

The invention of the sewing machine only began to take place during the 18th century, at a time when the clothing industry was trying to improve the production rate of the artisans on which it still relied. An important element was undeniably the **double-pointed needle** — in the form of a shuttle — with the eye in the centre, which was patented in 1755 by the American Charles T. Wiesenthal. His was a simple idea intended to avoid the seamstresses and embroiderers having to turn the needle round for every stitch on both sides of the cloth. The patent must have been little known, for a sewing machine project which the Englishman Thomas Saint patented in 1790 did not mention it. In theory the two main elements of the invention, the **movement** and the **needle** were ready; all that remained was for them to be united.

Forty years later, in 1830, Barthélemy Thimonnier, a humble tailor from Saint-Etienne, designed and made the first sewing machine. This was in fact a table on which a **hand-wheel** moved a connecting-rod, the up-and-down movement of which made the Wiesenthal needle rise and fall. The machine, which was made of wood, was rudimentary but it worked. Thimonnier obtained his letters of patent and although his machine had no means of pulling the cloth through and it required skill to achieve a regular stitch, in 1841 eighty models were in use in Paris for making army uniforms. They were competition for the seamstresses, however, and the unfortunate Thimonnier was nearly torn to pieces when a furious crowd attacked his shop and smashed up his machines. Unyielding, and no doubt remembering the fate reserved for **Jacquard's looms**, Thimonnier nevertheless improved his machine in two stages and in 1845 he protected it with patents both in England and the United States. The 1848 revolution ruined his business. His sewing machine, though greatly improved — it was thereafter made entirely of metal — aroused hardly any interest among the onlookers at the 1851 Great Exhibition in London. In 1857, when Thimonnier improved it for the last time, it seemed to be no more than just another dusty curiosity in the history of inventions.

Elsewhere it was not to be forgotten, for between 1832 and 1834 the American Walter Hunt had designed and made a more 'advanced' sewing machine than that of Thimonnier. The needle was fixed to an oscillating arm, forming a thread loop on the underside of the material, through which a **shuttle**, which was also swinging, passed a second thread. This made the **machine stitch**, using two threads, which was to be used in all machines. However, it was not Hunt but his compatriot Elias Howe who patented this type of machine in 1844, and was to have little success with it. Although he had sold his patent to an English corset maker, William Thomas, and he had actually been in person to London to promote his invention, Howe returned penniless to the United States five years later.

Thus in the middle of the 19th century, although it had been tested, the sewing machine did not interest very many people. The American Isaac Merritt Singer made some improvements to Howe's invention and patented it. Tremendous lawsuits followed and the publicity incurred must have alerted the public to the existence of the sewing machine.

In 1849, the same year that Howe had gone home discouraged to the United States, the American Allen B. Wilson improved the sewing machine yet again by introducing the **automatic feeding** of the

> Thimonnier's second machine could do 200 stitches a minute; the seamstresses' anger can therefore be imagined. His third could do **chain stitch**, whence the name given to it by Thimonnier and his colleague Magnin: *cousobrodeur* (embroiderer-sewer).

cloth, and in 1851 another American William O. Grover invented a machine which made the **double chain stitch**. When in 1857 a farmer from Virginia, James E.O. Gibbs, built a simple machine which did the **single chain stitch** and therefore enabled any housewife to sew the family's clothes mechanically, the sewing machine slowly began to climb the slope which led to it becoming both a classical household appliance and the basic equipment for the large-scale clothing industry.

Silk

Loui-tsé (?), 3rd millennium, BC

The Chinese were the first to understand that the average length of 1500 m of thread which made up the **bombyx cocoon** of the **mulberry worm** could be unwound and woven. The great age of the cultivation of the silkworm or **sericulture** in China is attested in numerous documents and no previous accounts of silk weaving have been found anywhere else. According to Chinese tradition it was Loui-Tsé, the wife of the mythical Asian emperor Houang-ti (Huangdi), who revealed its secret to her subjects in 2698 BC. This origin is not improbable, but it cannot be verified. The fact remains that the silk trade was an important source of revenue for China very early on. For centuries it remained mysterious, featuring on the list of the precious goods of the world. This trade created a steady flow of traffic and commercial trading across central Asia, the famous **Silk Route** by which many other kinds of merchandise and discovery were conveyed. The region which stretched beyond the Ganges was called the Land of the Silk or **Sericum** by the Romans.

It is interesting to note that in the 4th century BC the inhabitants of the Greek island of Kos themselves discovered how to weave silk, as reported by Aristotle, but the Romans knew nothing about the nature of silk (which they regarded as a kind of plant down), either because the weaving of it in Kos stopped between the 4th century BC and the beginning of our time, or because its secret was very well kept. It was only in 555 that the secret was revealed. Some Nestorian monks actually brought silkworm eggs back to the Mediterranean (supposedly hidden in hollow canes). **Cocooneries** were also founded during the second half of the 6th century in Athens, Thebes and Corinth, and the proliferation of mulberry woods earned the Peloponnese the name of **Morea**. In the 12th century, Roger of Normandy, the king of Sicily, imported some Hellenic craftsmen and founded a silk industry at Palermo. Thereafter the silk industry took root in the West.

> With the importation of silk to France being so expensive, the agronomist Olivier de Serre decided to create a French silk industry. He planned how to adapt the cultivation of the mulberry tree to the climate, and in 1599 he ordered 20 000 of these trees to be planted in the Garden of the Tuileries.

Spinning wheel

Anon, China, 11th century BC

The spinning wheel, which after the **distaff** (see p. 117) played an essential role in spinning until the invention of weaving looms, appeared in China in the 11th century BC. It was derived from the machines used for unwinding cocoons of silk, for the silk industry had existed in China since the 14th century BC. In the 1st century BC another Chinese invention, the **driving belt**, in addition to the development of cotton cultivation, probably gave rise to the pedal-powered spinning wheel. It was from this that the European spinning wheel, which appeared in the 13th century, was derived.

Stamping

Anon, Greece or China, 8th century BC; Vinci, c.1500; Cellini, c.1530; Philippe, 1835; Cavé, c.1836; Japy, 1838; Palmer, 1848

Stamping, or the use of pressure in the manufacture of metal coins, is almost as ancient as metal money itself (see p. 159). Anonymously invented, it dates back to Greece or China in the 8th century; the precise place has not been established. The pieces of money were made by inserting small ingots of metal, which had been heated beforehand, between two **hard metal moulds**, the top one of which was hammered down. This process was also used until the 19th century, in India for example. It is worth mentioning in passing the introduction of **molten moulding** by the Romans around the 2nd century BC. It is estimated to have been around 1500 that Leonardo da Vinci designed a **stamping press with a screw**. It is not known whether he built this but a version was made by Benvenuto Cellini for the purpose of stamping money, with the only difference being that it was a **lever press**. Stamping hardly progressed at all in method until the first half of the 19th century, when steam was mastered.

In 1835 the Parisian wheelmaker E. Philippe reached the next stage in the progress of stamping with his **bending machine**; this consisted of large wheels joined to cylinders between which he passed the bands to be bent, following a principle derived from **rolling metal**. The same process was then used to make the curves on the sides of boilers and water tanks. Around 1836 the Frenchman Cavé made a bending machine for stamping which was derived from the **power-hammer** (see p. 133): the sheet of metal to be stamped was first heated and then subjected to the pressure of a **chuck** in the desired shape, on to which the hammer fell. In 1838 the Swiss Louis Japy patented a press which went back to the principle of the lever for making copper and sheet-metal cooking utensils. From 1848 the Frenchman Jean Palmer submitted a series of patents which resulted in a method this time called two-stroke **stamping**, the first by lever, the second by power-hammer. Stamping provided the source of modern **sheet-metal manufacture** in the **car**, **maritime** and **aeronautical industries**.

Standardization

Gribeauval, 1776; Whitney, 1798; Taylor, 1881

Historical research has established that the first man to design and bring into effect the industrial mass production of parts in such a way that they were all identical, and therefore interchangeable, was the French general Jean-Baptiste de Gribeauval. Realizing that one of the biggest problems in military supply was the repair of damaged weapons, owing to the fact that they all had different parts and therefore did not allow the interchangeable use of spares, Gribeauval invented a method of **rationalization** of the manufacturing process in the arsenals, which became famous under the name of the '**Gribeauval system**'. It had many problems, one of the most important being the construction of **gauges** for the **measuring** of all the parts, a new system which introduced a totally different way of doing things from that of the craftsman. The Gribeauval system held the attention of many engineers across the world, and in 1798 the American Eli Whitney took it up and applied it again to the production of **hand weapons**, this time with more success. It was then used by Zachary Taylor, also a general (commanding the American armies in the War of Mexico); Taylor demanded of the firm Colt that the thousand revolvers which he had ordered from them be strictly identical. Colt employed Whitney's son, Eli Whitney, Jr, to devise a method of '**standardization**', but again this was a failure, with the parts being made by hand without machine tools. The first person who really seemed to have been able to apply the principle of chain production was the American clockmaker, Eli Terry, around the year 1860.

Standardization made a decisive step forward in 1881 with another Taylor, Frederick Winslow Taylor. He was a machine operator in the Midvale Steel Company and during an argument with the workers in the section of which he was in charge, he found it necessary to define an industrial working day. He then undertook a definition of the tasks attributed to each worker. This led him to his theory of **scientific management**, and to the distribution of the tasks to be done. The Gribeauval system thus entered into the industrial age, with the allocation of precise jobs going hand in hand with the necessity to produce standardized parts.

The two firms to be the first to achieve profitable standardization were Singer in 1876 and McCormick in the 1890s. Singer manufactured sewing machines and McCormick made harvesters.

Suspension bridge

Anon, China, around 25 BC

The first account of a suspension bridge dates back to the year 25 BC; it was found in a Chinese dynastic text which mentions a span of 15 m (in fact it consisted of several similar bridges which spanned the San-Ch'ih-p'an gorge in the Himalayas). However, a later account dating from 399 and describing the same region cites a span of 120 m. The bridges were made of bamboo. In 1638 Major Li Fang-hsien (Li Fangxian) built a suspension bridge made out of metal chains, with a span of 45 m. The first suspension bridge in the West was the Wynch bridge over the Tees, in England; it had a 20 m span.

The progress of the technology of building suspension bridges in the industrial age, using new materials, posed considerable problems, as attested by the building of the *Oakland Bay Bridge* in San Francisco in 1936, and the collapse of the *Tacoma Narrows* bridge in 1940. The large dimensions were in fact the cause of powerful distortions.

Weaving looms

Anon, Egypt, 4400 BC; anon, China, 2500 BC; Bouchon, 1725; Vaucanson, 1728; Kay, 1733; Cartwright, 1785; Jacquard, 1801

The oldest description of a loom is provided by a drawing found on some Egyptian pottery dating from 4400 BC. It was a very simple device consisting only of two parallel rollers which were to survive for a long time in weaving. One, called the **warp beam**, unwound the warp threads towards the weaver, while the other rolled up the cloth as it was made. The warp beam was steadied with weights and placed on trestles.

It was here that the fundamental invention in weaving appeared, that of the **heddle shaft** a set of two horizontal bars on the loom between which there were cords or **heddles**. These were regularly spaced and had the **warp threads** passing through them — each gap comprised an eyelet through which was threaded a warp thread.

There seems to have been only one heddle shaft on this Egyptian loom, which the weaver raised (or lowered) by hand. Assuming that it was the even threads that were threaded, this would have been, for example, to free them, in such a way that the layer of even threads and odd threads formed a dihedral angle through which the weft thread, perpendicular to the warp, was passed. A rod featured on the drawing could have been a rudimentary **reed** designed to pull together the **pick** — which is the name for a length of weft — or else to separate the even and odd threads — which would nowadays be called a **shed rod**.

Fragments of damask silk found in tombs dating from around 2500 BC in China imply that they had an improved version of the loom described above. The damask could not possibly have been made on the ancient loom, which was too rudimentary. It would normally have comprised two heddle shafts, one carrying the even threads, the other carrying the odd threads, which may have been fixed to frames so that they could be operated by pedals.

The fact remains that in Europe in the 13th century there were looms which had two heddle shafts and **pedals**. The heddles were mounted on frames, as was the batten, a kind of reed for pulling together the picks. The batten frame was attached to the actual uprights of the loom by pivots which allowed it to be pulled down by hand. The two essential improvements were firstly the setting of the heddles on frames operated by pedals, which freed the hands and meant that the weft thread on the shuttle could be moved very quickly, and secondly, the confined batten which could also be operated by hand in a single movement. It then became possible to increase the productivity of a loom considerably, and also, during subsequent centuries, to increase the variety of patterns. Thus in the Middle Ages there were looms with 25 or even a hundred

It was in mending Vaucanson's loom that Jacquard thought of the improvement which was to result in his own loom. Thereafter one worker alone could produce considerable quantities (up to 50 m) of cloth during the course of a working day. This made numerous workers redundant and Jacquard received a hostile welcome from the silk workers in Lyons who threw several of his looms into the Rhône.

Jacquard's loom

heddles, the latter being probably of Asian origin.

To do elaborate weaving, the warp threads on the warp beam followed a predetermined order and were arranged in groups: the blue threads, the yellow threads, etc. An apprentice would have perched on the loom to lift them by hand to allow the independent movement of the heddle shafts, which, being extended over the whole width of the loom, would not admit segmented separation. At the beginning of the 17th century a mechanical system was invented, apparently in Italy, which did the work of the apprentice, and much more reliably too; this system, which was similar to that of piano keys, raised different groups of threads and was controlled by a pedal.

In 1725 the Frenchman Basile Bouchon introduced the first system of **perforated cards** for the control of this mechanism. A roll of paper, perforated according to a pattern, unwound from a cylinder, which was pushed towards a **box of needles** controlling the hammers which selected the threads. If there was a hole, the needle entered it and the thread which it controlled did not move; if there was no hole, the needles were raised and went on to the threads. It was a reliable system but it too required an assistant; it was improved in 1728 both by increasing the number of threads and by the creation of an independent card for each pick, which meant that there was a series of perforated cards.

In 1728 another Frenchman, Jacques Vaucanson, the famous automaton manufacturer, modified the Bouchon system by placing the perforated card mechanism above the loom. This resulted in the threads which were to be detached from the warp layer, and which were therefore threaded in the needles, being raised directly, with the needles being attached to hooks which were lifted by a metal bar. It was a complex system and is unlikely to have been commercially successful. Its unquestionable merit lay in the fact that it paved the way for Jacquard's invention.

Between Vaucanson's loom and Jacquard's loom a significant improvement took place in the form of the **flying shuttle** which was invented in 1733 by the Englishman John Kay. Being a specialist in large-scale weaving, which suffered from the fact that the shuttle did not always reach the other side, or sometimes got mixed up with the warp threads, Kay designed a mechanism which had enough force to send the shuttle over the whole distance and then retrieved it using a string to which it was attached. In this way the imperfections in the weaving were eliminated, or at least reduced. When it was presented at the 1801 Universal Exhibition in Paris, the Jacquard loom was in fact an improved version of the Vaucanson loom. Its method was to raise the warp threads mechanically by the intervention of a system of strings which lifted the heddles in the same way, under the control of a set of needles. These were controlled by perforated cards which were adapted to a single prism and unwound continuously. The cards contained the design to be made. The control needles, which were fixed to rolled springs, made certain heddles rise when they entered the hole in the card.

The whole of this apparatus increased the height of the weaving loom itself and resulted in it being three times larger.

The first mechanical weaving, which was patented in 1785 by the Englishman Edmund Cartwright, was not very successful, but it did serve as a basis for the first effective mechanical looms. These were made in 1803 and 1813 by the Scot William Horrocks, and included essentially the automatic unwinding of the warp threads and the winding-up of the fabric on the beam. The **cast-iron frames** meant that the size of the looms could be considerably reduced. They ensured a much greater stability and consequently reliability too.

Measuring Instruments and Mathematics

One of the few areas in the history of inventions where continuity can be seen is that of instruments and techniques for measurement and observation. With the exception of the European Middle Ages, where virtually nothing in this domain was produced, curious minds have always been busy compiling an inventory of the visible world. Since the 4th millennium BC, time has been measured; since the 3rd, the Earth has been measured; in the 2nd, objects were weighed with growing precision; in the 1st, the Chinese invented the decimal system and Hipparchus used a theodolite to measure horizontal angles.

Mathematics is the mother of this movement, her offspring being mechanics, optics, physics, chemistry and statistics. The human mind develops a system of exact references which progressively drive out theories on the perception of the world and replace them with facts. No longer can one have the 'impression' that a certain point is further away or nearer than another, nor that the wind is blowing more or less strongly today than it did yesterday, nor too that the death rate has changed in such a way since such a date.

One of the most astonishing inventions before the end of the 19th century was that of the calculating machine. At first it seemed that it was intended only to allow calculations to be made more quickly, and the aim of the adding machine was to alleviate old Mr Pascal's arithmetical work; in fact, it was to alter the conception of logic, then psychology, and then the philosophy of knowledge. It eventually resulted in the invention of the clock. This substituted the objective notion of time for that of internal time or duration, which was founded by Bergson. Though a perfectly material measuring instrument, the clock is undoubtedly the only thing which has changed the human being from the inside, for in the 20th century the very organization of measurable time was, by means of automation, to penetrate the timing of the division of labour and of simultaneous industrial operations. The clock was thus to create impatience as well, and the strange feeling of boredom during a transatlantic crossing which lasts only seven hours in an aeroplane instead of the 2000 hours it took two centuries ago.

Anemometer

Anon, Persia, c.700; Alberti, c.1450; Hooke, 1644; Lind, 1775; Robinson, 1846; Dines, 1891

Attempts to measure the strength of the wind are relatively old. In the 1st century BC the Greeks used a primary device, consisting of a thin sheet of wood fixed vertically on a pivoting axle between two poles; the deflection of this indicated the **dynamic pressure** of the air. Around the 7th century the Persians measured the speed of the wind by the rotation of the blades on **windmills**, with the rotation of the mill being proportional to that of the blades. Such was the principle of the **vane anemometer** described by the Italian Leone Alberti around the year 1450, which was specifically for measuring the speed of the wind; Alberti did not definitely build this device and the credit for it goes to the English scientist Robert Hooke in 1644.

In an attempt to get rid of the effects caused by the friction of the blades on their axles, the Scot James Lind, who is known especially for having identified and remedied the causes of scurvy among sailors on long voyages, made a **pressure anemometer** in 1775. In this the dynamic force of the wind was exerted on a tube facing the flow which measured dynamic pressure, called a Pitot tube, after the Frenchman Henri Pitot by whom it was designed in 1750. Another, perpendicular tube measured the **static pressure**, and the tubes were linked to a **differential manometer**, which indicated the differential pressure due to the force of the wind. This type of anemometer was improved in 1891 by the Englishman W.H. Dines and it is still used. In 1846 the vane anemometer had been improved as well by the Irishman Thomas R. Robinson who made the **cup anemometer** which is still used today.

Balance

Anon, Egypt, 5000 and 2000 BC; anon, Campania or Great-Greece, 2nd century BC; da Vinci, 1480–1500; Roberval, 1669; Wyatt, 1744; Béranger, 1849

As an essential instrument in trading, the balance dates back to at least 5000 BC and to the origins of the Egyptian civilization (even perhaps of the Mesopotamian civilization as well). The first balances consisted of a **beam** attached by a string to a **column**, with **pans** at each end which were also hanging by strings. A **tare** would be placed in one pan and the goods to be weighed in the other. The **point of inertia**, when the two pans balanced, was judged by guesswork. Around 2000 BC the Egyptians improved this system in the following way: the beam had a longitudinal hole made in it through which a string was passed; this was taken to each end of the beam and knotted, and it supported the pans on each side, thus enabling the inevitable inequalities between the arms to be corrected much more easily. Around 1500 BC the Egyptians added a **lead wire** to their beam balances which hung from a bracket fixed to the column, below which there was a triangular marker fixed to the beam. This enabled both the horizontality of the beam and the exactitude of the equilibrium to be verified. Around the 1st century the Romans simplified this model by fixing the beam to the column using a **transversal pivot** and then later by placing the ends of this pivot on **forks**. The indicator of the point of inertia, which was to become the **needle**, was fixed to the fork and corresponded to a mark on the beam.

It is not known at exactly which stage in his life Leonardo da Vinci invented the first **graduated dial balance**, the main principle of which was adapted from the old beam balances, but which included a remarkable innovation. Whilst the other balances of the time did not actually measure the weights which were placed on them but were instead restricted to establishing the equivalences between predetermined **weights** and objects, this one indicated the weight of the suspended object; it was therefore the first **automatic balance**. Everything points to the fact that he made it himself, or else he built several versions of it, certainly between 1480 and 1500. The instrument had a **semicircular graduated scale** fixed at the centre to a column. Its beam had only one arm, which extended the diameter of the dial; the other arm was replaced by a specific weight to which a needle was attached, and this followed a course in front of the dial. Any pressure exerted on the pan suspended from the single shaft of the beam would be indicated therefore by a movement of the needle across the graduations of the dial. It is not known whether it was due to the fact that it was not published or to mere indifference that this balance, which was three and a half centuries ahead of known techniques of weighing, passed more or less unnoticed.

It was only in 1669 that the French mathematician Gilles Personnier de Roberval fundamentally modified the design of the balance with his '**static enigma**'. In this he introduced the principle of the two **coupled beams**. These were joined by the arms with four pivots to form a rectangle or virtual parallelogram. Spars were soldered to the centre of each shaft and had weights suspended from them. The enigma lay in the fact that even when weights were suspended at unequal distances, the two beams remained horizontal. The equilibrium was upset only when a pressure was exerted at the end of the upper or main beam, that is, when a weight was placed on one of the pans of the balance. The phenomenon was to be explained in 1821 by the mathematician Louis Poinsot with the **couple theory**. In the meantime Roberval's balance, which was derived from the 'enigma', was successful owing to the fact that it was transportable. It inspired a great many versions to be made, and intrigued innumerable scientists such as Alembert. One of the best-known versions was that of the Frenchman Joseph Béranger who patented it in London in 1849. In short, Béranger split up the lower beam into two independent beams, thus refining the static equilibrium obtained by Roberval.

The third type of balance which came into use from the 19th century is paradoxically very old; it was called the **steelyard** balance and seems to have been invented during the 2nd century BC in Campania or in Great-Greece. It was actually a variation on the antique balance described above, of which a great many versions existed, and evolved from Rome to India and Denmark; today they are known under the names of **hook steelyard**, **butcher's steelyard** and **open-air steelyard** for weighing barrels, packages, etc. It consists of a dissymetrical beam with a hook suspended at the short end and a fixed weight sliding along the long end which is toothed and graduated. The distance of this weight in relation to the suspension point of the beam — a ring — determines the resultant of the forces exerted on the arms of the

A great many types of balances were invented in the 19th century and became the object of even more variations and improvements; it would take a treatise to discuss them all. Among the most outstanding there are the **weighbridges**, a **Duchesne system** inspired distantly by the Wyatt balance and designed for weighing livestock and vehicles; the **dial balance**, a Duchesne system; the **Merrick balance for weighing towed objects**, a **Viguier system**; the **Coulomb balance**, also called the **torsion balance**, which was for measuring the **magnetic forces** of magnets by the torsion of copper or silver wires; the **Eötvös balance**, a type of torsion balance comprising a wire terminated with two spheres and in which the irregularity of the oscillations indicates **gravitational differences in land masses**; the **Curie balance**, which has shock absorbers for very precise measurements; the **electromagnetic balance**, which enables the **intensity of current** to be measured by the attraction of a coil on a magnet; and **chemical, pneumatic and optical microbalances**.

beam; the further away the weight moves from the suspension point, the more pressure it exerts and the more its relative value increases. If for example the weight was 10 kg, it would be brought as close as possible to the suspension point so that it balanced a mass of 10 kg; as the mass increased in weight so the weight would be pushed back towards the other end.

In 1744 the Englishman John Wyatt invented the **lever bascule** which did not seem to meet with the success it deserved until the 1830s. The principle behind it was that of a floating platform on to which the weight to be measured was placed. This was usually something reasonably heavy, such as a loaded cart. The platform exerted its pressure on the ends of two opposing V-shaped levers; these in turn transmitted the pressure to a central lever; the distance of this from its point of inertia indicated the weight which was required to compensate the pressure exerted by the platform and therefore the weight of the object.

Barometer

Torricelli, 1643

In 1640 the Italian physicist Gasparo Berti fixed a lead pipe, the lower end of which was submerged in a tub of water, to the wall of his house. He filled the pipe with water and sealed its top end. To his surprise he found that it was impossible to make the level rise above 10 m by adding water. His brilliant compatriot Galileo experienced the same puzzlement when pumping water from a well; the column of water never rose above 10.3 m. Neither man understood the reason for this limit, which was established by a third Italian Torricelli, who was the first to recognize the part played by the pressure of the air. In 1643 he made the first **barometer**, or instrument for measuring atmospheric pressure. He replaced the water with mercury, the column of which at normal atmospheric pressure measures 73 cm. The term 'barometer' was coined by the Frenchman Edme Mariotte. In 1665 the Englishman Robert Hooke added a **needle dial** to Torricelli's graduated barometer. This was controlled by a wire attached to a float at the surface of the mercury and allowed more precise measurements. In 1844 the Frenchman Lucien Vidie invented the **aneroid barometer**; the needle of this is controlled by the deformations of a metal capsule inside which a vacuum has been created.

Perier measuring the height of the Torricelli tube on top of Puy de Dôme. Experimenting with atmospheric pressure. Engraving by Yan Dargent.

Calculating machine

Anon, China, 9th century BC; Napier, 1617; Gunter, 1620; Pascal, 1642; Leibniz, 1694; Thomas, 1820; Babbage, 1822

The history of the calculating machine is about as complex as the machine itself, firstly because several of its stages consist of mutual loans, with one inventor borrowing a part or a principle from another which he then altered, and also because several of these machines had different aims and therefore carried out different operations. For this reason there is a case for speaking of several distinct machines rather than just one machine.

The **abacus** invented in China in the 9th century BC is unanimously recognized as the first of all calculating 'machines'. Clearly it was not a machine since it was not mechanical, but it already had the big advantage of enabling the materialization of arithmetical calculations by the displacement of balls strung on axes in a frame, with each row of balls representing different countable units, a million, a hundred thousand, ten thousand, a thousand, a hundred, one, a tenth, a hundreth, etc.

For a long time it was thought that it had no successor before modern times, which we will place at around 1500. Now this has been put into question since the discovery in 1947 of a puzzling bronze box which was fished out of the sea off the coast of the Greek island of Anticythera. This contained several interlocking cogwheels of different sizes which were mounted on axes and is known as the '**Anticythera clock**'. This device calculated the positions of the seven planets known at the time and therefore constituted, in the 2nd century BC, an astonishing precursor to the first calculating machines as well as to **astronomical clocks** (see p. 152).

The fact remains that the next stage was not reached until 1617, with the Scot John Napier's '**knucklebones**', a mobile **logarithms** table consisting of segmented sticks which allowed mathematical operations which would otherwise have taken many long hours to be carried out quickly. This instrument operated on the common basis of logarithms which were invented by Napier and the Englishman Henry Briggs. It was in fact a **multiplier**. It was very successful, particularly among astronomers and, of course, mathematicians.

In 1620 the English astronomer and mathematician Edmund Gunter improved Napier's invention by fixing the 'knucklebones' on to a surface, consequently making the first **slide rule**. This too was refined, by his compatriot Henry Leadbetter.

In 1642, at the age of 19, Blaise Pascal invented a machine comprising a clever **gearing** system which allowed **additions** and **subtractions** to be carried out by the simple manipulation of six wheels on the lid of a small oblong box. The sums could be seen in small windows placed above the wheels. The young Pascal's invention could however be regarded as an adaptation of the mechanisms used in clockmaking. This was undoubtedly the case with the one made by his English predecessor William Schickard, which was a less elaborate machine of the same type.

The German Gottfried Wilhelm Leibniz, who admired Pascal, took up his invention, which had already had considerable success; in 1671 he began to improve it in order to make it capable of doing **multiplications** (by successive additions) and **divisions** (by successive subtractions) in addition to the two other basic mathematical operations. He finished his project and his machine was exhibited in 1694 at the Royal Society in London. It could also extract **square roots**.

These machines were known as **digital adding machines** because they carried out arithmetical operations by counting **whole numbers**.

For more than a century people endeavoured to make commercial machines based on the inventions by Pascal and Leibniz, but without much success. It was only in 1820 that the first results of this research appeared, starting with the extremely successful and commercial **arith-**

mometer made by the Alsatian Xavier Thomas, and then with the improved versions which came to light half a century later made by the Englishmen F.T. Baldwin and W.T. Odhner.

Meanwhile the English mathematician Charles Babbage tackled the problem of speed with, as his objective, a machine which combined **arithmetical functions**, like those of Pascal and Leibniz, with **logical functions**; this machine would therefore make decisions according to its results. Later on, Babbage also incorporated into his project the capacity to compare quantities and to follow predetermined instructions, and then to inject the results obtained to control a second series of operations. It was an incredible project for the time since it implied the combining of two types of machine, the **digital** and the **analytical**, at the same time as associating two types of **function**. In order to do it Babbage made full use of the **punched cards** technique for the input of data, which was then widely used in **weaving looms** (see p. 140); the other operations were effected by means of gearing systems and levers.

Babbage, who paid homage to the genius of Vaucanson, the inventor of punched cards, was at the crossroads of automation, later to become computing. A true genius himself, the only reason that he did not complete his research was a lack of financial capital and sufficient technology; on his own he was unable to make the excessively complex apparatus necessary for his project; the limited support given to him by the British government was inadequate. Babbage exhibited a rudimentary prototype of his first machine in 1822, which he had called a '**differential machine**' because it calculated and printed tables of functions using predetermined differential techniques, but he died without having been able to complete his second prototype which was a **universal digital and analytical calculator**. The first prototype, which followed a fixed frame of operations or, as it is called in modern computing, a **fixed program**, could not in fact claim to be universal. His machine was not actually built until the 1970s, by IBM (although an adaptation of it was made in 1860 by the Swedish firm Scheutz).

Babbage was familiar with the work being done by his compatriot George Boole, the inventor of **symbolic logic**, who reduced all the logical relations to simple expressions such as ET and OU. He recognized the genius of Boole, whose algebra now serves as a basis for all electronic calculation and which enabled all mathematical functions to be expressed on a **binary base**, 0 and 1. It was in drawing inspiration from Boole, the author of the first treatise on **analytical transformation**, that Babbage tackled his huge project of the universal calculator.

Calculator program

Byron, 1835

It was the poet Byron's daughter Ada, countess of Lovelace, who in 1835, when she was only 20 years old, invented the first **sub-program** for an **arithmetic calculating machine**, which in this case was that of the Englishman Charles Babbage with whom she was acquainted. This sub-program, which was made on **punched cards**, controlled a certain number of specific parts in Babbage's analytical machine (see above), allowing certain arithmetical operations to be carried out automatically, thus freeing the operator from introducing the same instructions into the machine each time. Ada Byron was also the first to devise a **complete program** for Babbage's machine, still using punched cards. This remarkable invention could not come to a satisfactory conclusion any more than Babbage's machine, which remained at the embryonic stage until the inventor's death.

Calliper rule

Anon, China, 9 BC; Vernier, 1631

The discovery of a calliper rule, an instrument for measuring the diameter or the thickness of an object, bearing its date of manufacture — which was equivalent to the year 9 BC — has allowed the Chinese to be attributed with this invention. The inventor remains unknown. The instrument was made of wood and was used as a model for the realization of one in bronze.

The calliper rule was introduced in Europe only in 1631, by the Frenchman Pierre Vernier, although Leonardo da Vinci seems to have designed one about 150 years previously. It was improved by the Englishman William Gascoigne in 1638 for the purpose of measuring celestial objects, and hence it earned the name of **micrometer**.

Cartography

Anon, Mesopotamia, 3rd millennium BC; Eratosthenes, 3rd century BC; Hipparchus, c.130 BC; Mercator, 1569; Picard, 1669–1670; Tissot, 1881 etc

Cartography is without doubt one of the most complex areas in invention history, for it implies in the first place a certain amount of **topographical**, **geographical** and **geodesical information**, and in the second place, a **mathematical** method of writing down spherical or hemispherical representations on a level surface.

Cartography is very old; we now have maps which date back to the 3rd millennium BC. Indeed, cartography must have been indispensable at a very early date to the administrations of kingdoms and then to the despots, for the purpose of establishing frontiers and, finally, to travellers. At first it consisted of empirical representations, which were inevitably inaccurate, both because the distance calculations themselves were incorrect and then because the representations were plane as though the Earth were flat.

Being great travellers, the Greeks were among the first to become interested in general and not cadastral cartography, the latter being practised by the rest of the world. In the 6th century BC, Miletus, in Asia Minor, was the centre of cosmological and geographical speculations. The first cartographer worthy of the title was Thales's pupil Anaximander, who designed the first geographical map, followed by Hecatus in the 5th century who gave the first representation of the Earth: a flat disc at the centre of which was Greece. In this respect he cannot be said to have been an inventor; he was just a precursor. Herodotus, a great traveller who belonged to the next generation, enriched the geographical knowledge of his time considerably. This he did, for example, by questioning the existence of a hyperborean ocean and being the first to realize that the Caspian Sea was an entire sea and not a gulf of the hyperborean sea. Xenophon's reports on the campaign of the Ten Thousand, the campaigns of Alexandria, the knowledge of Megasthenes, who was Greek ambassador to the Hindu court of Chandragupta, and many other pieces of information caused considerable changes in the representation of the Earth as it had been until then. In the 3rd century before Christ one of the most astonishing geniuses of antiquity, Eratosthenes, was in a position to conclude that the Earth was spherical, and succeeded in the amazing feat of calculating almost exactly the circumference of the meridian to be 252 000 stages, which is nearly 40 000 km (39 690 to be precise). Eratosthenes can therefore be considered as the real founder of cartography. Even though the map made by him, for lack of more extensive information, had only three obviously very roughly drawn continents, Europe, Asia and Africa, it was the very first to feature **longitudes** and **latitudes**, and the first also to provide a **spherical representation** of the Earth. To do this Eratosthenes had based his work on the information collected by a great 4th-century BC navigator, Pytheas, who had crossed the Columns of Hercules — now the strait of Gibraltar — and had gone around the British Isles, making precise recordings of the **declinations of the Sun** in the latitudes of the Northern hemisphere. Using the distances indicated by Pytheas and his solar declinations, Eratosthenes did not go far wrong — about 300 km — in his estimation of the **Earth's circumference** (which besides has perhaps changed since then).

However in 130 BC, another great scientist in antiquity, the mathematician and astronomer Hipparchus, made a judicious objection to Eratosthenes's technique, for it offered a global vision of the Earth according to the distance calculations of a small number of places. He suggested that these calculations could be inexact — which was true — with Eratosthenes having used an empirical method to create a precise theory.

Hipparchus proposed a radical reform of cartography. His method consisted of dividing the globe into a longitudinal and latitudinal grid of 360° and determining the situation of different places on the basis of **astronomical observations**. Hipparchus therefore invented a **geodesical system** to determine the longitudes according to the observation of **eclipses**.

It has been suggested that Hipparchus could not have applied his method strictly, owing to a lack of precise observation instruments; this has not been verified. In the 20th century off the Greek island of Anticythera, the remains of an astonishingly precise **astronomical clock** were found, the workings of which are still being determined. The fact remains that Hipparchus is universally recognized as the founder of scientific cartography.

A short while after the decline of Alexandria as the scientific centre of antiquity, which is thought to have occurred after the 2nd century, an increasingly dense shadow fell over cartography. The Romans let this science lapse. They needed maps for their administration — the Empire covered the whole of the Mediterranean world and a large part of the Near and Middle East — so they turned to the method of **equivalent projections**: this consisted of writing down exactly the distance between two points, rather like spreading out all the pages of an atlas edge to edge without taking into account the sphericity of the Earth. This is the method used in the old **administration maps** in modern times, and is absolutely valid for **small-scale maps**. Their maps were essentially those of servicemen and administrators rather than scientists. Marcus Agrippa took 20 years to make a huge map of the Empire under the orders of the emperor Augustus which recorded the roads, bridges, frontiers, and imperial and senatorial provinces, etc, and was etched on a marble wall near the Colosseum in Rome. A later copy of it, called Peutinger's map, has been preserved; it includes excessive irregularities and hardly seems to correspond to any **geodesical system** at all; both the Mediterranean and the Black Seas, for example, are so elongated that they look like canals. Rome, however, did not worry about either the

A spherical surface can be projected on to any geometrical surface: cylinder, cone, polyhedron, etc. Each method of projection is linked to a specific technique. A great number of modes of projection do exist, the exposition of which would require a treatise. The most common are the **conformal**, **equal-area** and **equidistant** modes. The first, also known as **orthomorphic**, is used for large-scale maps and imposes a transformation of the geodesical curves. The second, also known as **equivalent** or **authalic** and used in administration maps and small-scale tourist maps, treats the sections of the sphere as though they were flat. The third is used in scientific research and comprises large deformations. Contemporary aeronautical maps are equal-area Mercator projections.

exactitude of proportions or mathematical aesthetics in this domain.

Cartography in the Latin world underwent a full and dramatic degeneration during the 3rd century. The Christian apologist Lactance rejected the idea of the sphericity of the Earth and thereafter the Christian cartographers drew it as a flat, O-shaped disc, inside which the T or *tau* of the Cross was inscribed. It had Europe on the left surmounted by Paradise, Africa on the right, and the Nile in the middle — and the bar of the T represented by the Mediterranean. The admirable work done by Eratosthenes, Hipparchus and, later, Ptolemy, seemed likely to disappear altogether under the attack of these so-called mythical cartographers; it was to be preserved only by the Arabs, and it was an Arab, Idrisi, who made a silver planisphere (1154) for Roger of Normandy, king of Sicily, which was to perpetrate and enrich the Greeks' knowledge. It was another Arab called Birini who, in the 11th century, quite rightly rejected the medieval fabrications about the southern lands which were to be linked to Africa. Indeed the Arabs, also navigators, had land and sea links with Asia; they possessed elements on the geodesy of the orient and Asia by means of the trade of silk, ivory, spices and stones. In the 16th century the Turkish admiral Ahmet Muhieddin, better known under the name of Piri Reis, made maps which have given rise to ardent controversy in the 20th century. Dating from 1513 these maps delineate the coasts of North America, South America, the Antarctic, the Caribbean and the islands in the North Atlantic with a wealth of detail that corresponds neither to the discovery of North America by Christopher Columbus in 1492, nor to that of South America by Amerigo Vespucci between 1499 and 1502. These points which are still vague are essential, not only with regard to geography and history, but also to cartography. Indeed at the time, the civilized world laboriously pieced together the knowledge of the Greek cartographers; Eratosthenes's calculations had been forgotten and neither the **Earth's circumference** nor the **length of the meridian** were known. Previous knowledge, which was perhaps the work of Arab, or else African, navigators (Bénin's Empire would therefore seem to have had connections with Central America in the 14th century), or maybe Phoenicians (for it seems also that the Phoenicians reached what is now Brazil during the 2nd or 3rd centuries before Christ), may have enabled Reis to draw up these maps which, even if they are not exact attest to unexplained cartographical knowledge.

For whatever reason, from the 15th century the Arab influence provoked a renaissance of cartography in Catalonia and Italy. The discovery in 1400 of Ptolemy's *Geography* — which was much less precise than that of Eratosthenes and Hipparchus — inspired Western cartography to spread anew. Ptolemy's map was redrawn in Bologna in 1475 using a **conical projection**, and then again in 1482 by the German Donis Nicolaus using a **trapezoidal projection**.

From this time onwards the Western cartographers, who had returned to a spherical representation of the Earth, began to take note of the fundamental problem in cartography which arose from **two-dimensional geometrical geodesy**: creating a plane representation of something which does not have an applicable surface. This representation would inevitably be distorted owing to the projected distances being increasingly longer or shorter as the distance increased from the contact point O between the model and the volume of reference (cone, cylinder, polyhedron, etc). Thus if a globe of the Earth is placed in a transparent cylinder with the same diameter, the only point of contact will be at the equator. The distances between the various points on the equator will be reflected correctly, for example from Gabon to the mouth of the Amazon and to the Galapagos Islands. But when a **direct projection** is used, the distances between the latitudes above and below the equator will diminish as they get further away from the equator towards the Poles. In this way England's surface area will be considerably reduced in relation to her size. If the proportions are to be respected, a **transverse projection** must be used, but if this is realized continuously a different type of distortion takes place: the distances between the longitudes are exaggerated in relation to an equatorial scale of reference. Moreover, if these maps

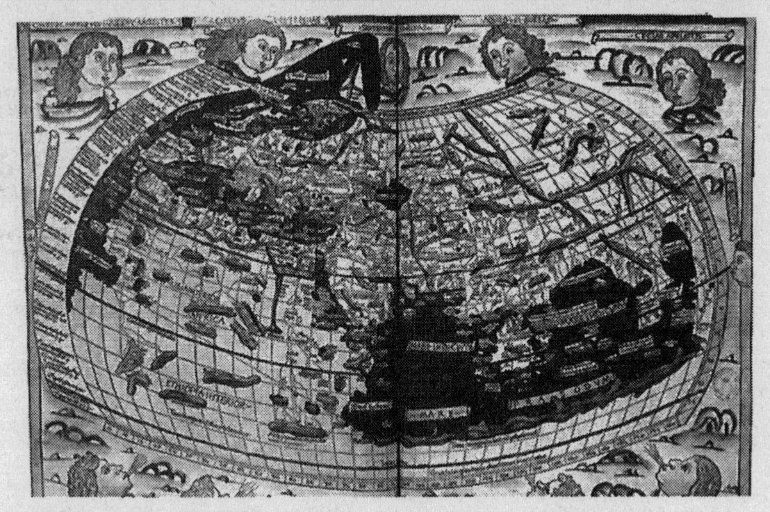

***World map by Ptolemy**, in a printed edition of* Geography *translated into Latin, Ulm, Léonard Hol, 1482*

are to be used by travellers and navigators they must indicate the distances from one point to another on a **fixed scale**.

From the 14th century onwards, navigators used **portulans**. These were marine maps which indicated the distances from a fixed point to several other points following a fixed scale. They nevertheless had an important fault: they did not show the convergence of the meridians and therefore deformed the distances between the high latitudes. In 1569 the Fleming Gerardus Mercator, a Latin pseudonym for Gerhard Kremer, published a map of the world for use by navigators which stood out because it resolved the contradiction which had made geometrical projections unacceptable in marine maps. Realized by **cylindrical projection** it in fact indicated — empirically — the straight line on which the compass had to be centred in order to go from one point to another, a line which maintained the same angle in relation to all the meridians. It inaugurated 'Mercator' projections which were to be endlessly improved throughout the ensuing centuries, resulting in **standard cylindrical projections**.

It is appropriate to mention nevertheless that 'Mercator's' projection was invented in the 5th century in China, but for the purpose of representing the celestial sphere and not the Earth. A map dating from 940 shows the sky divided in this way into 27 meridians. Later on, this method of projection may have influenced the European cartographers.

Mercator's representation was still a long way from being geographically correct, for the knowledge of the time was still thoroughly engrained with mistakes concerning the longitudes and the latitudes, and all the more because no one really knew the Earth's dimensions. In 1606 the Dutchman Willebrord Snell undertook to overcome this deficiency by being the first to measure a **meridian arc**; his work was followed up by the Englishman Richard Norwood in 1635. It was only in 1669–1670, however, that the Frenchman Jean Picard carried out this measurement in a scientific fashion. This was not actual cartography, but **geodesy**; nevertheless, without the progress made in geodesy, cartography itself would not have progressed. It was other geodesical progress accomplished during the missions of Maupertuis to Lapland and of La Condamine to Peru that was eventually to give a precise picture of the object of cartography, that is, the shape and dimensions of the Earth, a globe slightly flattened at the poles.

In 1881 the Frenchman A. Tissot made great progress in cartography with his *Report on the representation of surfaces and cartographical projections* using **calculus**: if a surface were projected on to a plan following the **equivalence method**, that is, by establishing a point-to-point correspondence, an infinitesimal circle would become an ellipse. It therefore became possible to calculate the proportions of any ellipse on any projection, and, from the orientation of its large and small axis, to establish the orientation of any point. On an equivalent projection for example, the surface area of an ellipse would be equivalent to that of the projected circle, but its elongation and its axes would vary according to each point.

Once all the geographical and geodesical knowledge of the Earth had been assimilated, the application of **solid geometry** was to enable the realization of maps which were correct in all respects.

Clock

Anon, Egypt, 3500 BC; anon, Egypt, 1400 BC; Berosus, 300 BC; anon, Alexandria, 1st century; I-Hsing (Yi Xing), 725; Gerbert, c.950; Henlein, c.1500; Huygens, 1656; Hooke, 1660; Bain, 1840; Popp and Resch, 1864(?)–1877

The **measurement of time** is one of humankind's oldest concerns. The word 'horology' comes from the Greek **hôrologion** ('that which tells the time'), but the clock as a specific instrument did not appear until the 7th century in China; however it is difficult to ignore the clock's precursors. The first system for measuring time was the **sundial** or **gnomon** which was already in use in 3500 BC and perhaps even before then. It consisted of a horizontal surface with a vertical pillar stuck into it, the shadow of which indicated the movement of the Sun. In the 8th century BC the gnomon was to be improved to become the proper sundial; on its base it had six divisions each corresponding to an hour, and at one end there was a vertical arm. In the morning this arm was pointed towards the East and in the afternoon the dial was turned so that the arm pointed towards the West. From the 14th century BC the Egyptians had also used **water clocks** or **clepsydras** which had the advantage of continuing to indicate the time after sunset, although their imprecision was a problem. The clepsydras consisted of vessels with holes in their bases out of which water would drip regularly. Since the vessel was graded inside with marks corresponding to the passing of an hour, the time could be read off. The length of the hours changed according to the length of the day, which meant that the summer hours were longer than the winter hours and that their number was not constant. Furthermore, the dripping of the water was subject to variations resulting from the temperature. The clepsydra became widely used in the Graeco-Roman world at an unknown time and it too was improved. First of all the Romans improved the flow of water by installing larger tanks, since the flow would be more regular under a greater pressure. Then they invented an original mechanism for reading the hour, consisting of a **float** equipped with a toothed stem connected to a **pawl wheel**; the lowering of the float engaged the wheel which itself had a needle to designate the corresponding hour on a graduated dial. This mechanism was already in use by the 1st century and so the principle of the modern clock, the reading of a dial, can be said to date back to this time.

In the meantime the sundial was improved again, notably by the Babylonian astronomer Berosus in the 3rd century BC. Berosus had transformed it into a square block with a hemisphere hollowed out of its upper surface. This had the point of a style reaching to its exact centre. The shadow of the style made an arc which would have had a much more precise indication than that of the previous dials, provided that the graduation took into account the **precession of the Earth's axis of rotation** which was unknown at the time, and that the device had metrical dimensions, which were also unknown. Half a century later Apollonius of Perga made a dial which was even more advanced, with the **hour lines** inscribed on a surface of **conic section**, and Ptolemy even improved the principle of the sundial using the **analemma**, where the shadow of the style was projected on to inclined instead of horizontal surfaces. The fact remained that these measuring methods did not enable the difference between **average midday** and **true midday** to be established, and they could only be used during daytime. Even though they were still used until the 10th century and maybe even later, especially in the Mediterranean world, interest in the clepsydra continued to grow.

This invention arrived in China, where a portable clepsydra was made. In this the water was replaced by **mercury** and it was used to measure short time periods. This was more an adaptation than an actual invention. In 725 the Chinese monk and mathematician I-Hsing (Yi Xing) invented and made the first known **mechanical clock**, which was derived from the clepsydra. Its driving power was still hydraulic, but it was more precise than the holed vessel. It consisted of a **wheel** with strictly identical **paddles** into which the water flowed; each time a paddle filled up, it caused a rotation of a 36th of an arc. A huge gearing system (the clock was about 10 m high) of admirable precision caused the rotation of a celestial sphere around which the Sun and the Moon were represented so that there was one complete rotation of the Sun every 365 days and one complete rotation of the Moon after slightly more than 29 days. The level of a part representing the horizon also enabled the exact hours of sunrise and sunset to be determined, as well as the dates of the new and full moons, and of course, the hours and quarter-hours which were announced by bells and drumbeats ... A later version made in 906 by Chang Hsu-hsün (Zhang Xuxun) showed the movements of the five planets, the Polar star and the Great Bear. A third version, designed this time by Su Sung (Su Song) and built in 1092, showed the movements of the stars and certain special days and hours; it was installed at Kaifeng, the capital of the Songs, and it functioned until the fall of that dynasty in 1126.

The principle of Chinese clocks probably arrived in the West in the form of descriptions or drawings. It was definitely in the West however that the first **weight clocks** appeared. Their invention is attributed to Gerbert, a Benedictine monk born in Aquitaine who was elected as pope in 999 under the name of Sylvester II. The principle was simple, and was clearly more reliable than running water (the clocks described above were incapacitated in freezing conditions). A weight was suspended by a cord from a lighter counterweight which represented the weight of the cord; this was wound round a cylinder, the regular rotation of which was controlled by a cogwheel; in turn the rotation of this was regulated by an axle comprising two opposing cogs which

> Gerbert, Pope Sylvester II, had such remarkable mathematical, mechanical and cosmographical knowledge that he earned a sulphurous reputation for wizardry, with some people accusing him of practising sorcery. Gerbert had learnt his sciences from the Arabs in Spain.

> In the 13th century the town of Nuremberg specialized in the making of small domestic clocks, to which it owed a large part of its prosperity. At the time and until the 15th century the mechanisms of clocks were uncovered. It seems to have been only gradually, beginning in England, that they began to be provided with protective covers.

engaged each other alternately by successive pivoting. In this way the first model of the **escapement system** was made, which was later to be improved in the form of **cylinder escapement** (with the axle meanwhile having been enlarged and modified). It was an invention of appreciable importance, since for the first time clocks could really function autonomously. The first clocks nevertheless had neither **dials** nor **needles**, but a **hammer stamp** which at given intervals would activate a ringer, which would then ring the bells. There are two versions of this type of clock remaining to this day which were built for Salisbury Cathedral in England and for Rouen, in 1386 and 1389 respectively. The one in Rouen has the unprecedented characteristic of sounding the quarter-hour. At the same time smaller clocks were made for domestic use.

Middle Ages pendulum. Wood engraving in German incunabulum, 1490

The next improvement arrived around 1500 and was the work of the German Peter Henlein. He made the first **domestic spring drive clock**, with a horizontal dial on top and a single hand, the one which indicated the hours. Around 1656 the Dutchman Christiaan Huygens invented the **pendulum clock** which had proper **cylinder escapement**. In 1660 the Englishman Robert Hooke completed the first **anchor escapement** which had the particular advantage of being able to transmit some of its energy to the pendulum. Between then and the mid-19th century, several improvements were made to the weights and the pendulum with the aim of eliminating the irregularities caused by the thermal expansions and contractions of the parts.

In 1840 the Frenchman Alexander Bain made the first **electric clock** (where the electric energy served only to give a regular force to certain cogwheels), which nevertheless still had the weight and the pendulum. In 1864 the Austrians Popp and Resch produced the first clock to be powered by compressed air or **pneumatic clock**, but it did not become viable until 1877. A set of pneumatic clocks was started in Paris in 1880.

Compass

Anon, 1st century

The origin of this instrument, which is essential to the discovery of the world and on which all geographical maps are based, seems to go back to **geomancy** and to a divination device described in the year 83 in a Chinese book, the *Louen Heng*. As shown in a bas-relief dated 114 which is in Zurich museum, it was a shaped spoon made of **magnetized iron ore**, or **magnetite**, which was placed on a plate of polished bronze to see in which direction the handle would point.

This instrument was therefore most probably based on a discovery, that of magnetite. It is not known exactly when the discovery gave rise to the invention, but it is known to have taken place in China, for **floating or pivoting needle compasses** have been found from the 9th century, when the instrument did not exist in Europe. The

Houa Chou (Hua Zhou) reports that in 940 the engineer Chen Koua (Chen Gua) even knew how to make **artificial magnets** by cooling a steel bar when it was in a north-south facing position.

The first known use in **maritime navigation** was reported in 1187 by the Englishman Alexander Neckham who described 'a pointer carried on board which enables a course to be followed even when the Polar star is hidden by clouds'. Until 1700 **compasses** were fixed and mounted on maps, where the North was indicated by a fleur-de-lis. Subsequently, **liquid compasses** were made, where the needle was placed in a metal box filled with distilled water. Since the needles tended to become quickly demagnetized, in 1745 the English engineer Gowan Knight invented a technique for magnetizing hard steels on a long-term basis. In 1876 Sir William Thomson, later to be Lord Kelvin, introduced a **dry compass**, the binnacle of which was itself mounted on a pivot and included a system for the correction of the errors caused by the ship's own magnetism. This type of compass disappeared when greater speeds made the needle vibrate excessively and they were replaced by liquid compasses once more.

Decimal system

Anon, China, 14th century BC

Although the invention of **zero** has been successively attributed to the Arabs and to the Hindus in historical times, there is good reason to think that it was invented at the same time as the decimal system by the Chinese mathematicians in the 14th century before Christ. Calculation was carried out then using rods on a squared board and the Chinese seem to have been the first to represent the number 10 by a single rod followed by an empty square. The number 11 was represented by two single rods in two adjacent squares, which could have been confused with the number 2, which was indicated by two rods in the same square. The number 3 was 'written' by placing a rod in three adjacent squares.

The first proof of the use of the decimal system was found in a Spanish manuscript dating from 976.

The empty square sufficed to represent zero, therefore, and even though it is not certain whether the Chinese thought of a specific sign for zero before they devised the other signs (these consisted of circles and have been found in Cambodia and Sumatra, dating back to 683, about two centuries before Gwalior's famous inscription in India which dates from 870), it is certainly they who invented the concept.

Differential calculus, integral calculus

Newton, 1665; Leibniz, 1676

The closely linked differential calculus and integral calculus are major inventions in the history of the human brain and can be understood in the following way (even if one does not have advanced knowledge of the subject). In **algebra** the common operations are **addition, subtraction, multiplication, division, raising to the powers of numbers** and **taking the roots of numbers**. In calculus one has recourse to an additional operation which consists of establishing a **limit**. In **plane**

geometry this operation is called upon, for example, in considering the surface of a circle as the limit of the surfaces of inscribed regular polygons with $4n$ sides which have no limit; in other words, if the number of polygons increases, their surfaces decrease, and if the number decreases, their surfaces increase.

There are two determinant limits in algebra: one is the **derivation of a function**, the other is the **defined integral**. Thus, if y is a function of x, $y=\int(x)$ and it follows from the relative increase in y in relation to x, that is, the derivation of y in relation to x, that if x undergoes an increase $\triangle x$, x becomes $x+\triangle x$ and y becomes $\int(x+\triangle x)$. y has then undergone an increase which is $\triangle y$, its new value minus its initial value, which is written as

$$\triangle y=\int(x+\triangle x)-\int x$$

To estimate briefly the relative variation of $\triangle y$ in relation to $\triangle x$, the result is assumed to be the increase in y divided by the increase in x, that is, $\triangle y/\triangle x$. This estimation is too brief, however, and suffices only to define a value of the quotient. On the strength of this the increase $\triangle x$ is reduced indefinitely until it approaches zero and the variation of the relation is followed. Of course, as $\triangle x$ approaches zero, $\triangle y$ approaches it too, but it would be perfunctory to note that their relation is written down as 0/0. In fact, if one supposes that their relation $\triangle x=0.1$ and that $\triangle y=0.41$ (for evidently their values are not the same), their relation is found to be $\triangle y/\triangle x=4.1$. In **analytical geometry** (see p. 160), one considers that when $\triangle x$ and $\triangle y$ are both very small, their relation is related to the tangent to the given section of a curve. This very brief demonstration suffices to show the idea central to differential calculus. If it is pursued further, the limit of the relation $\triangle y/\triangle x$ can be established (if $x=2$, the angle of the tangent is 4). This was a new way of approaching analytical geometry which was to have a great influence, particularly on **mechanics**.

The limit implied by integral calculus is the **limit of a sum**, for example of surfaces or lengths of arcs, or surfaces determined by certain curves. The following demonstration of it can be given: if $\int(x)$ is a single function of x defined in all the points of an interval $a\leqslant x\leqslant b$, it is a case of finding the points x_i in this interval such that $x=a$, then $x_1<x_2$ ending up with $<xn=b$. Then at each interval $\triangle x_i$ a point t_i must be found. Finally, the sum of these intervals must be established. The limit of this sum, as each interval $\triangle x_i$ approaches zero, that is, as the number of subdivisions in intervals increases, is designated as the definite angle, $\int_a^b\int(x)dx$, if this exists; it does exist if $\int$ really is a continuous function of $a\leqslant x\leqslant b$ . . .

The significance of integral calculus was to spread to all sciences, from physics and astronomy to thermodynamics and mechanics.

The invention of differential calculus and integral calculus will always remain a debatable point in the history of inventions. There is no doubt that they were both conceived by Sir Isaac Newton at the time of the bubonic plague which raged in London in 1664–1665 and which forced him into reclusion in Woolsthorpe, since Cambridge University was closed. This is demonstrated by the scientist's manuscripts.

The same invention, almost certainly later, is credited to another great scientist, the German Gottfried Wilhelm Leibniz who would have achieved it in 1676. The curious thing is that until 1672 Leibniz did not know much about mathematics; he was initiated into it that year by the brilliant Christiaan Huygens. Leibniz's was undeniably a matchless brain, but it must be noted that in 1673 he went to London, where his English friends had told him that the quadrature of the hyperbola by Mercator had led Newton to devise differential calculus (about which the English scientist had not yet published anything). Leibniz therefore knew at least that Newton had invented differential calculus, and perhaps he knew more, for the plague epidemic had come to an end and Leibniz had many English friends who undoubtedly enthused to him about Newton's invention.

The fact remains that in 1677 he published his **fundamental theorem** on differential calculus, twelve years after Newton's invention. Being a diplomat by trade and endowed with an often acute opportunism, Leibniz managed to imbue his invention with considerable import; differential calculus and integral calculus quickly became known as 'Leibnizian calculus'. Leibniz is therefore attributed with the invention of the differential $dy = \int '(n)dx$, the quotient dn/dx of which gives the derivation, while Newton, according to some, was connected only with the derivation. On seeing this Newton became angry, published his own work and accused Leibniz of plagiary. More than three centuries later the whole business has still not been resolved, owing just as much to Newton's undeniable precedence as to what must be called, for want of a better word, Leibniz's dishonesty.

His **method of infinite series**, for example, bears a curious resemblance to that which was previously invented by the Scottish mathematician James Gregory, and moreover it is common knowledge that Leibniz 'borrowed' copious extracts from Baruch de Spinoza's as yet unpublished *Ethics*, which he 'forgot' to mention ... Until further information is available it will not be possible to rule out the possibility that Leibniz used what he had learned of Newton's work on differential calculus and integral calculus in order to advance his own reputation.

Dividers

Hindley, c.1739; Ramsden, 1760–1773

The development of **astronomy** at the beginning of the 18th century created a growing demand for **measuring instruments** and for instruments that were more and more precise. The particular problem with these instruments, which were handmade, was the difficulty of the **division** of the limbs to be graduated, a problem which had both a theoretical and a practical aspect. The first was the **geometrical division** of the arc of the circle, and the second was the **engraving** of the marks. The most scrupulous craftsmen made the divisions using a compass, while the others applied a universal model. What was needed was an instrument which would mechanically divide a circle into 360 strictly identical degrees. The first to come near to a solution was the Englishman Henry Hindley in about 1739, who adapted the **machine for shaping clock parts**; this was an **endless screw**, the thread of which corresponded to a determined circumference and number of teeth (360 in this case). Each contact of the thread with the tangent of the circle caused a notch to be made. The instrument made noticeable mistakes.

From 1760 to 1773 the Englishman Jesse Ramsden, one of the most famous manufacturers of his time, designed and built a machine for the making of the parts which had previously required special machines. In principle this was a **dividing lathe** consisting of a circular toothed platform and a deeply threaded screw. The instrument to be divided was placed on the platform. An operating pedal moved the platform via a crank, and at the same time it lowered the tracer or **tracing-awl** which made a mark on the metal being divided. The precision obtained was in the order of a sixth of a degree and the division time reduced to half an hour. This invention was immensely significant since it lowered the cost price of astronomical measuring instruments, such as **octants** and **sextants**, and at the same time allowed a great many more of them to be made.

Geometries

Anon, Babylon, beginning of the 2nd millennium BC; anon, Egypt, 1700 or 1550 BC; Thales, 7th–6th Century BC; Hippias of Elis, 4th Century BC; Euclid, 330–320 BC; Archimedes, c. 260 BC; Apollonius of Perga, c.245 BC; Descartes, 1637; Fermat, 1679; Monge, 1780; Lobatchevski, 1826; Gauss, 1827; Plücker, 1828–1865; Bolyai, 1831–1833; Möbius, 1837; Riemann, 1854

In geometry and in mathematics the capacity for abstraction is manifested to its highest level. It is incorrect however to evoke a unified geometry, for there are actually several geometries, which for convenience are divided into Euclidian and non-Euclidian. In fact, the history of geometry is mixed up with that of a succession of inventions.

The significance of geometry is considerable, for it dominated **architecture** and then the **knowledge of the Earth**, then that of the Universe and the **symbolic representation of non-perceptible systems**. It is possible here to retrace its development only in a very succinct way, and the names cited above cannot give complete expression to a whole string of inventions which has justified extensive works.

Geometry was born a very long time ago, for **cadastral reasons**. The oldest origins date back to the Babylonians at the beginning of the 2nd millennium BC; they worked out a system for measuring their territories which was based on **units of simple surface, circumference and square** which must have been easily calculated. Besides, according to its Greek roots the very term geometry designates the measurement of the Earth. Babylonian geometry was empirical and approximate; thus it gave π the value of 3. It is nevertheless from it that we have inherited the division of the circle into 360 equal parts or degrees, the division of the degree into 60 minutes and that of the minute into 60 seconds — a system which, it is interesting to note in passing, is duodecimal, even though it co-exists universally with the decimal system (see p. 155). The Hebrews do not seem to have altered the elements of Babylonian geometry, and it was the Egyptians who made the next step forward. The papyrus of Ahmes, which according to different sources dates from 1700 or 1550, enumerates the means of calculating the **areas of a square** and **a rectangle** as well as **an isosceles triangle** and **a trapezium**. It was the Egyptians too who advanced the definition of π, as attested by the method written in the same papyrus for calculating the **area of a circle:** in order to do this it was necessary to subtract 1/9 of the diameteer and work out the square of the remaining 8/9, whence the result that $\pi = \pm 3.16$. Another almost contemporary papyrus gives the exact method of calculating the **volume of a truncated pyramid**.

The above is all related to empirical geometry. Abstraction reached its first state of definition with the legendary character Thales, one of the seven sages of ancient Greek philosophy, who delineated the following **five geometrical theorems**: the bisection of the circle by its diameter; the equality of the angles of an isosceles triangle; the equality of the angles formed by the intersection of two straight lines; the theorem stating that an angle inscribed inside a semicircle is a right angle, and the theorem stating that it is possible to draw a triangle if the length of its base and the angles of the base are stipulated. It is this last theorem which would have enabled Thales to calculate from the shore how far out to sea a ship was situated.

A new step was taken by another legendary character, Pythagoras, a philosopher who founded a school and who apparently lived in the first third of the 6th century BC. It is difficult to know exactly which of his contributions to geometry were his very own and which came from his school. The fact remains that he was the first to prove that the sum of the angles of a triangle will always be equal to that of two right angles, and above all he would have established how to calculate the areas. Pythagoras would at least have started on the **properties of parallelograms** and is also associated with the invention of the formula of the square of the hypotenuse of a right-angled triangle. Finally he is attributed with the invention of the method of constructing the **five regular polyhedrons** (tetrahedron, cube, octahedron, dodecahedron and icosahedron), although it is likely to have been his pupil Théatète who invented the octahedron and the icosahedron.

With the taste for abstraction generating one for geometry, the Sophists of Athens set down, in the stride of Pythagorean speculations, the three famous construction problems of the **quadrature of the circle**, the **trisection of the angle**, and the **doubling of the cube**. These sophisms are sometimes represented as intellectual challenges and gratuitous games for the mind. However they had the great advantage of alerting the geometers to the limits which they had imposed upon themselves and which subsequently would be associated with those of Euclidian geometry. Consisting of intellectual challenges which were impossible to solve with the then traditional geometrical instruments, the compass and the ruler, they led the geometers to pass to a higher level of abstraction. This was the case, moreoever, when in the 4th century BC Hippias of Elis invented a method to resolve the quadrature of the circle using a curve of his own creation, the **quadratrix**, and when towards the middle of the 3rd century, Nicodemes invented another curve, the **conchoid**, which enabled both the trisection of the angle and the doubling of the cube, the latter operation being summarized algebraically by the formula $X = \sqrt[3]{2}S$, with S representing the edge of the cube.

However much they pay homage to Euclid, contemporary geometricians and mathematicians nevertheless emphasize the precursory genius of the Greek non-Euclidians, whose inventions prefigured those which twenty centuries later were to give rise to non-Euclidian geometry and which were already touching on the most advanced views of modern physical theory, geometry being associated to physics in the representation of space. Thus the quadratrix of Hippias of Elis already involved the idea of a number of infinite points, which developed into the modern notion of **continuum**. 'Gaps' such as these in traditional Greek logic were placed in a vein which already existed and which had, for example, been illustrated by Zeno of Elea in the 5th century in his famous paradoxes on motion. For Zeno, the arrow fired by the archer did not fly because its path consisted of an infinite succession of moments of rest, and the famous Achilles could never capture the tortoise because he ran much faster than it did. Until the beginning of the 20th century the sophisms and the paradoxes of the first geometers and thinkers were held with suspicion as 'curiosities', or as demonstrations of the traps which were the risk of the over-rigorous exercising of logic. In fact they had the benefit of pulling geometry away from its previous pragmatism.

Between 330 and 320 BC, in that incomparable centre of intellectual fermentation, Alexandria in Egypt, Euclid wrote his *Elements* of geometry which consisted of thirteen books. The first six and last three books were devoted to solid and plane geometry, the seventh, eighth and ninth to arithmetic and the tenth to irrational numbers. Strictly speaking, it did not constitute a work of invention, not entirely anyway, but rather it was a logical compilation of the immense heritage bequeathed by Euclid's predecessors. In it there were 35 **definitions**, six **postulates** and ten **axioms**. Euclid appeared to be a rather restrictive organizer. Even though he too, if only apparently, disengaged geometry from pragmatism (which his successors Archimedes and Apollonius of Perga would hardly ever do), he carefully avoided the notions of **infinitely small** and **infinitely large**, the exposition of which had been to the credit of Zeno and Hippias. However this every rudimentary account of an immense work, which itself concludes the three prodigious centuries of Greek geometry, must not be concluded on a negative note: Euclid invented a formal method, **logic**, which had lasting success. Until the beginning of the 20th century Euclid was to be used almost exclusively as an introduction in the teaching of geometry throughout the entire world.

Euclid had only two significant successors: Archimedes and Apollonius of Pergamum. Around 260 BC the former invented (or discovered) the necessary formula for the calculation of the area of a circle, $3\frac{1}{7} > \pi > 3\frac{10}{71}$, as well as the formulae for the calculation of the surface area of a sphere and a cylinder. At the same time the **application of areas** enabled Apollonius of Pergamum to construct the **three conical sections**.

With the exception of a few later secondary geniuses, such as the above-mentioned Nicomedes, Diocles, who invented the **cissoid**, and Pappus, Greek geometry had by this time completed its own parabola.

Taken up again by the Arabs after the fall of Alexandria in 642, it was transferred to the West and significantly criticized and enriched from the 12th century, but it did not really become widespread until after the fall of Byzantium in the 15th century. Shortly afterwards the Renaissance began in Europe: Europe was ripe for the assimilation of Greek geometry after the long sleep of the Middle Ages. The first development added by the geometers was **projective geometry** which was first of all to be used in **cartography** (see p. 148); it then founded **perspective**, or the representation of spatial figures in a plane from a certain viewpoint. Perspective would itself be enriched by the notion of **infinity** invented in 1639 by the Frenchman Gérard Desargues, who was a friend of the famous Marin Mersenne. Desargues himself was also to influence Blaise Pascal.

René Descartes and Pierre de Fermat were the two first great successors of the Greeks in that they were the inventors of the very important **analytical geometry**. It is assuredly appropriate to pay homage in this domain to the German Johannes Kepler who re-introduced the notions of the infinitely small and the infinitely large after some twenty centuries, and defined the circle as a polygon with an infinite number of edges. The Italian Bonaventura Cavalieri must be mentioned too, for he took up the geometry of solids where the Greeks had left it, as should his compatriots Evangelista Torricelli and Vincenzo Viviani, and the Frenchmen Gilles de Roberval and Blaise Pascal, who contributed to the development of modern scientific geometry. Their works, however, never attained the importance of those by Descartes and Fermat. With *La Geometrie* published in 1637 and *Ad locos planos et solidos isagoge* published in 1679, Descartes and Fermat respectively were the first to establish the **formal junction of geometry and mathematics** which had been outlined by Hippias of Elis and Nicomedes. They set out the two fundamental ideas of analytical geometry, which are the localization of the points of a figure by the use of coordinates and the representation of a curve or a surface by an equation comprising two or three variables. The former idea actually goes back to Apollonius of Pergamum, but the latter was revolutionary, for a little less than 250 years later it led to the first calculations of **general relativity**. In *La Geometrie*, which is not an exposition in the modern sense, but an account which is almost familiar in the style in which it tackles problems and resolves them, Descartes resolves in particular the **problem of Pappus**, after the Greek geometer who was the first to state it: given various straight lines to a point, find the position of a point such that a perpendicular drawn from this point to these lines makes the product of some of them have a given relation to the product of the others. Descartes realized that the problem should be understood thus: find the equation which reveals all the characteristics of the desired figure (which is a curve). He then tackled the problem with the supposition that it had been resolved; the desired curve was therefore considered to have been found and Descartes reasoned on finding out the information from it. The method owed its name to this kind of geometry, which proceeds by analysis and which enables problems of Euclidian geometry to be resolved just as well as non-Euclidian problems. It is a method which has astounded all those who have analyzed it, by its sheer power and universality. A great mathematician such as the Austrian Ludwig Boltzmann was later to write that it sometimes seemed more clever than the man who had invented it.

Descartes in fact founded modern geometry, rejecting the old classification of curves as **plane curves** (those which could be drawn using a ruler and compass), **solid curves** (which were in fact conical sections) and **linear curves** (such as the conchoid). He reduced these to two groups, those which he called 'geometrical' and which were in fact **algebraic**, and those which he called **mechanical**. However he shares the credit for having created analytical geometry with Fermat, because (but this was only learned later) Fermat was the first to establish that the equation $\int(x,y)=0$ represents a curve in the plane x, y, which constitutes the basic principle of analytical geometry. Fermat, who was adviser to the Parliament of Toulouse and not really a professional mathematician, did not publish his work. If it is Descartes's genius that prevails, strictly within the domain of geometry, it is because he created a convenient method; Fermat's right to glory was due to his having been the progenitor of **differential calculus** (see p. 155). Other mathematicians who developed differential calculus and integral calculus, were the German, Gottfried Wilhelm Leibniz, who invented the terms **abscissa** and **ordinate**, and the Englishman Isaac Newton.

Thereafter geometry diverged. In the classical line, rather abusively known as Euclidian, the Germans Augustus Möbius and Julius Plücker introduced the notion of **duality**, which they had invented independently. This notion is based on the fact that it is possible to give two interpretations to the same equation, for example $ux+vy+1=0$. One can in fact maintain that this equation comprises two variables, on the one hand x and y representing coordinates of points and u and v being constants, and on the other hand u and v representing lines and x and y constants, which means that in the first case the equation defines a straight line, and in the second, a point. The two mathematicians therefore abandoned the restricted notion of **cartesian coordinates** to replace it with that of **homogeneous coordinates**. This invention paved the lofty path to 'pure geometry', which differs from the **descriptive geometry** which was developed by Gaspard Monge (one of the founders of the polytechnic) from 1780, and for practical purposes, by his pupil Jean Victor Poncelet from 1822. Numerous geometers, of which Monge was the first, and including Möbius and Plücker, endeavoured to combine the two geometries, the analytical and the synthetic, the latter having been developed to near-perfection by Michel Chasles in France and K.G.C. von Staudt in Germany.

In the meantime, however, a major event had taken place: the invention of non-Euclidian geometry. Its inventor is traditionally recognized as the German Carl Friedrich Gauss, and its date of origin as 1827, that of the publication of *General research on curved surfaces*. In reality non-Euclidian geometry had had a long period of gestation dating at least from 1621. Indeed, in that year the Englishman Sir Henry Savile called the **fifth postulate of Euclid** into question. The postulate states: 'If a straight line falling across two straight lines makes interior angles on the same side which are smaller than the two right angles, the two lines elongated to infinity will meet on the side that the two angles are smaller than two right angles'. This postulate, he observed, is neither demonstrable nor undemonstrable; it cannot be proved. In 1773 the Italian Girolamo Saccheri tackled the problem in Euclid's defence, or so he thought, but his conclusions were false. However, in endeavouring to construct a **reductio ad absurdum** Saccheri actually outlined non-Euclidian geometry. His demonstration was as follows: if the perpendiculars AC and BD are raised on a straight line AB, and C and D are joined by a straight line, then it can be deduced that the angles ACD and BDC are equal; they may be right angles, but one can also work on the hypothesis that they are obtuse or acute. In fact Saccheri used the reductio ad absurdum here to approach the question of parallels, such as had been stated by Euclid: through a given point only one parallel can lead to a given straight line, which did not satisfy anyone.

Gauss did not tackle the question of parallels; it was his successors who took it up again. For his part, Gauss ratified the doubt of the validity of 'this strange geometry', or non-Euclidian geometry, which includes no contradictions. Among other novelties, in

1830 he introduced the notion of **variable space**, which is implied in his formula for the circumference of a circle of radius r, which is $\pi k(e^{r/k} - e^{r/k})$, where k is a constant which depends on the nature of the space. There is already the feeling that relativity was no more than three-quarters of a century away.

Non-Euclidian geometry was in the air of the time, for apparently without knowing about Gauss's work, between 1826 and 1840 the Russian Nicolas Lobatchevski 'corrected' Euclid's theories in the same way. For him, two parallel lines could be taken from a given straight line, each one of which would rejoin this line in infinity. It followed from this that a straight line had two points in infinity and that it made equal acute angles with its parallels, and that when the radius of a sphere became infinite, its radii would become parallel ... Without knowing about Lobatchevski's work, in 1823 the Hungarian János Bolyai arrived at more or less similar conclusions on his own (but he did not publish them until 1831, with his *Theory of Parallels* being published in 1840). These revolutionary works, which instituted the notion of the **curvature of space** — and, in short, meant that all that had to be added to end up with relativity was the measurement of time — passed by almost unnoticed. Their essence, however, imposed itself more and more, for the German Bernhard Riemann reinvented their conclusions in 1854 without knowing about Lobatchevski's and Bolyai's publications.

Hygrometer

Vinci, c.1500; Ferdinand II of Tuscany, 1650; Saussure, 1783; Boeckmann, 1802; Döbereiner, 1802; Regnault, 1845

The idea of measuring the **humidity** of the air preoccupied scientists for a long time because it could facilitate **weather forecasting**, but it seems that the first to have made an instrument for the purpose was Leonardo da Vinci around the year 1500. Precedence is sometimes conceded to the German cardinal Nicola de Cusa who may have thought of it half a century before Vinci, but this is not certain. Anyway, the technique used by both of them was the same; it consisted of measuring continuously the weight of a ball of wool; the ball weighed more when it was damp. The second type of hygrometer in scientific history was not much more 'scientific' than the first: it was proposed in 1783 by the Swiss Horace Benedict de Saussure and was based on the human hair's property of getting longer as it absorbs moisture. This idea is said to have been conceived already by Duke Ferdinand II of Tuscany who was interested in atmospheric measuring instruments. **Hair hygrometers** have been made again in the 20th century.

In 1802 the German C.W. Boeckmann thought of a different method which consisted of placing two thermometers side by side. One of these had its bulb exposed to the air whilst that of the other was wrapped in cotton; with the cotton absorbing the moisture, it followed that its temperature was lower so that the bigger the difference between the temperatures recorded on the two thermometers, the higher the humidity.

> Since the end of the 19th century, five different types of hygrometer have been recorded: the **evaporation** or Boeckmann-Babinet type; the **condensation** or Döbereiner-Regnault type; the **hair** or Saussure type, also known as hygroscopic; the **chemical** type, which is based on the reducing properties of humidity on certain substances, the weight of which varies according to the absorbed humidity; and the **electric** type, which measures the electric resistance of certain substances, such as lithium chloride, which varies according to the moisture they have absorbed.

This was basically a variation on Vinci's hygrometer which in 1824 the Frenchman Jacques Babinet took up again in his own name and renamed '**psychrometer**'. In 1802 the German J.W. Döbereiner devised a third method which was tested by the Frenchman Henri Victor Regnault in 1845: it consisted of cooling a highly polished silver cylinder and measuring the temperature at which the condensation of atmospheric water vapour took place on its surface, which indicated the **condensation point**. The lower the condensation point, the lower the relative humidity of the atmosphere.

Metal money

Prehistory, era unspecified

Material symbols of goods exchanged like money in transactions between individuals or from an individual to a group have existed since very early in prehistory. They are attested by the bronze and iron **arrowheads** which the inhabitants of Gaul used for this purpose several thousand years before Christ. This type of money survived for a long time, for large quantities have been found in tombs consecrated by Pheidon, king of Argos in the 8th century BC. It was called **obeliskos**, hence the term for a type of old French money, **obole**. Other forms of non-metallic money were used in Africa and Oceania, as well as in South America, such as the feathers and beaks of birds, shells and mats. These preceded the arrival of the first travellers by many years, which would suggest that the very concept of money existed long before the discovery of metals. The rolls of bird feathers used as money in Papua New Guinea until the beginning of the 20th century relate to a practice which could date back to three, four or even five thousand years before Christ, if not more. In the beginning these types of money were almost certainly used just as much for barter as for practical purposes. In the same way that the bronze arrow was in itself a precious object, (since it was less likely to break than a flint head and it was therefore possible to get it back from the corpse of the animal — or person — that had been shot), the feathers of birds and the shells of certain rare species had an ornamental value that meant that they were valued goods. Money was therefore used as if it were an object, a custom which survived until the beginning of the 20th century in some tribes in the north of Africa, where the women adorned themselves with pieces of gold. Until the 17th century in China a certain shell, **Epitonium pretiosum**, was so highly sought after that copies of it were made out of porcelain (and are now much more valuable than the originals). Money can therefore be said to be an extremely ancient collective invention.

> Whereas the Chinese used proper money from the 8th century BC, the ancient kingdoms in the Near East, the Sumerians, Babylonians, Assyrians and Hittites, did not have it until the 6th century. The first Persian coins seem to date back to 546 BC, the date when Cyrus conquered Lydia and consequently learned about the stamping of coins (see p. 136).

The recognized bases of barter were established according to custom. The value of a macaw feather, for example, depended on local conventions. These bases then became more precise. Until the end of the 5th century in Rome, fines were levied in the form of heads of livestock, but as it was not always easy to transport livestock from one place to another symbols of them, pieces of bronze imprinted with the head of a cow, were exchanged instead. As pastoral communities changed to become farming communites, which were less inclined to barter and more to trade, the use of these symbols spread. The herd, **pecus**, was then bought and sold on a financial basis.

In Egypt, where the transition took place

very early, gold **ingots** appeared around the year 1500 BC. Their weight was certified by the government of the kingdom and represented the maximum exchange values. Apparently beginning in Egypt, therefore, the weight of the metal attested to the financial and consequently fiduciary value of the symbol. The Egyptian practice cannot be claimed to set the example, however, for the Spartans used money made of iron in this way, as did the ancient Bretons, according to Julius Caesar. The Egyptians, thinking of foreign money as ingots, would cut them up into pieces; hence Greek coins which have been cut into two or four have been found in Egyptological excavations.

This practice derived from the fact that the first pieces of money were not regular; the same weight did not look the same from one piece to the next. The first to decide to normalize the weights were the Greeks and the Hindus simultaneously in the 3rd century. It was then that metal money in the modern sense was created. The coins were produced and guaranteed their value by the central power of the kingdom. It is worth noting here that this value included a significant margin of convention, for electrum, an alloy of gold and silver, did not always contain the same percentages of one or the other metal.

From this time onwards money lost the intrinsic value it had had when it consisted of arrow-heads or bird feathers, and since it was no more than a symbol, forgery began; false money is known to have been used in China in the 5th Century BC.

Meteorology

Fitzroy, Le Verrier, Smithsonian Institution, c.1840

Attempts to predict the weather are as old as humanity, but the first attempt at systematic international meteorology was only made possible with the establishment of **telegraph networks**, around 1840. It was the work of the Englishman Robert Fitzroy, the Frenchman Urbain Le Verrier, and the Smithsonian Institution in Washington. It concentrated on the **forecasting of major storms**. Around 1850 this rudimentary network had spread to numerous countries. At the beginning of the 20th century it was extended again with the advent of **wireless telegraphy**.

Metre

Mouton, 1670; Picard, 1671; Condorcet, Delambre, Méchain, 1792–1799

The metre, the basis of the **metrical system**, which is itself derived from the **decimal system**, is an invention which is both logical and singular. It is logical because it is inspired by a universal length of the Earth, that of **longitude**, and singular because it is exclusively French. It was first proposed in 1670 by Gabriel Mouton, vicar of the Saint-Paul church in Lyon, on the basis of a minute of the circumference of the Earth. The following year Jean Picard and other amateur astronomers took up the proposition, which still had no real consequences.

In 1792, after long deliberations, Marie Jean Antoine Caritat, marquis of Condorcet, and his colleagues from the Academy of Sciences, decided that the metre should be a **ten-millionth of a quarter of the Earth's circumference**; the decision was taken in agreement with the physicists Antoine Laurent de Lavoisier and Charles de Borda for the purpose of offering humankind a **universal measurement of length**. All that remained to establish a standard metre was to determine the exact length of the **meridian**. This was calculated in 1693 by the naturalized French Italian Giovanni Domenico Cassini on the basis of the geodesical observations of Jean Richter at Cayenne, and it was recalculated in 1771 by his grandson César Francois Cassini using more precise instruments. The assumption that the Earth was slightly flattened at the Poles required another verification which was entrusted to the astronomers Jean-Baptiste Delambre and Pierre Méchain in 1792; the undertaking, which comprised as much surveying as astronomical work, was firstly to establish the exact distance between two points on the same meridian in a given unit of length, and then to convert it into metres. The chosen points were Dunkirk in the north and Montjuich, near Barcelona, which are both 2° East in longitude.

It was an enormously useful undertaking because measurements varied so much from country to country. By way of example, the **drachma**, a weight measurement, was equal to 3.906 g in Holland, but 3.21 g in Turkey and 3.2 g in Greece. The Italian **fathom** measured 0.7 m and the French fathom, 1.62 m. The English **gallon** was worth 20% more than the American gallon. The Czechs measured in **latro** (1.917 m) but the **loket** measured 0.593 m in Prague, 0.594 m in Moravia and 0.579 m in Silesium. To add to the confusion, the Danish **pot** was 0.966 litres and the Swiss pot, 1.5 l, but the Belgian pot was only 0.51, while the Norwegian pot was 0.965 l.

These are not the only examples of the same term designating very different values: the Russian **vedro** equalled 12.33 l; the Bulgarian vedro, 12.8 l. Even though the English and the American **yard** were identical (0.9144 m), the Mexican yard measured only 0.838 m. To this very day the **oke**, which in Turkey is a kilogram, changes its value in Bulgaria where it can be used to measure liquids (1.28 l) or solids (1.282 kg). One of the most striking examples is that of the ounce: the old French **ounce** weighed 30.59 g, whereas the Dutch ounce was equal to 100 g. Complications in European and international trading can be easily imagined, as can the inevitable misunderstandings or trickery which were the scourge of commerce until the institution of the metrical system.

Delambre and Méchain did not finish their research until 1799, after a great many hitches which can be easily explained by the political events of the time. The metre was equal to 3 feet, 11 lines and 296/1000 of the toise of Peru. The work had enabled a valuable precision to be obtained: the theoretical metre calculated on the basis of Cassini's calculations was 145/1000 shorter in length than the toise of Peru. The standard metre was cast in platinum, because this metal has a low coefficient of expansion in heat.

On 20 May, 1875, an international treatise created the **International bureau of weights and measures** which endeavoured to improve the precision of the standard metre; it was recalculated following new work in **astronomical photometry** and recast, this time in an alloy of 90% platinum and 10% iridium maintained at 0 °C, despite the fact that it was identical to the metre recorded by Delambre and Méchain.

The definition of the unit of length led to that of a unit of weight, the **gram**, which is equal to 1 cm^3 of water at the temperature of maximum density, or 4°C. There was also a definition of a unit of liquid measure, the **litre**, equal to a cube with sides of 10 cm or **decimetre cube**. Finally there was the definition of a unit of area measurement or **are**, 100 m^2, and of volume, the **stere** or **cubic metre**. After becoming law in the United States in 1866, and in Great Britain in 1897, the metrical system existed alongside the old systems of measurement in these countries until 1973.

Modern calendar

Denis the Younger, 525

Every civilization and culture in the past has had its own system of calculating time, which has usually been based on a more or less exact estimation of the **astronomical year**, 365 and a quarter days. However, none of them used the same landmarks in time. By way of example, if the universal calendar had not been instituted, the year 1990 for the Buddhists would be the year 2444, dated from the death of Buddha; for the Muslims it would be the year 1278, dated from the hegira of Mohammed in 622; for the Jews, the year 5661, dated supposedly from the beginning of creation. Long before modern times these disparities caused all the more problems, for some peoples counted in **lunar years** of 13 months, and others in **solar years**. Furthermore, the start of the solar year was said by some, such as the English, to be 25 March, by others to be the spring equinox on 21 March, and by yet others, including the Romans, to be 1 January.

In 525 the Scythian monk or abbot Denis the Younger, also known by his Latin name Dionysius Exiguus, undertook (with the intervention of the pope John I) to propose a coherent calendar to Christendom. This was a vast undertaking, carried out in the guise of the organization of the liturgical year, given that Christendom was divided by the schism of the Church of Constantinople and that the ruler of Italy, Theodoric, of whom the pope was vassal, disputed the spiritual primacy with the Roman emperors of the orient, firstly Theodosius II, and then Justinian. A unifying calendar proposed, or imposed, by Rome therefore smacked strongly of a political venture. Until then, in fact, the Roman Christians and the Eastern Christians celebrated Easter on different dates.

For the Roman Christians the date from which Easter was calculated had been fixed according to a calendar which began with the emperor Diocletian — an absurd situation, since this emperor persecuted the Christians. Denis the Younger undertook in his calendar to make the Christian era begin with the birth of Jesus Christ, which he said was 25 December in the 753rd year after the founding of Rome. It was he who suggested to John I the writing of AD for **anno Domini** after the official dates of Christendom, in order to characterize the dating system more clearly.

This custom lasted for nearly half a century, when in 1583 Pope Gregory XIII was to go back reluctantly to the scholars' argumentation according to which it was not logical to make a year begin six days before it ended, for the Roman or Julian calendar was still in use in numerous regions of Europe, such as Italy. He therefore introduced a reformed Gregorian calendar, following which the year began on **1 January**. This caused a new outcry among the Christians of the orient, who had still not adopted Denis the Younger's reform. England continued to begin the year on 25 March as before, through hostility towards the papist calendar; the Gregorian calendar was only adopted there in 1751, and the English colonies in America did not adopt it until several years later. It was not adopted in China until 1911, and in the USSR until 1919.

Although he must be given the credit for establishing the starting date for a universal calendar, Denis the Younger made two mistakes in taking 25 December, 753 as the birth date of Jesus Christ. The first was a result of him being unaware this date, four days after the **winter solstice**, had been chosen by the 1st-century church because of its symbolic nature for the many peoples who had continued in the tradition of celebrating it; in the Mithraic religion, for example, December 25th is the date of the annual renaissance of the god Mithras after his victory over the powers of darkness. His second mistake, if one refers to the gospels themselves, was that Jesus was born during the reign of Herod the Great, who died in 4 BC. It has in fact been proven that the census commanded by Rome, owing to which Jesus was registered at Bethlehem, took place in 7 BC.

Photographic lens

Wollaston, 1812; Chevalier, 1821; Petzval, 1841

The photographic lens has the peculiar characteristic of having been invented before **photography**. The first lens ever made, which was later used in the first photographic instruments, was invented by the Englishman William Wollaston in 1812. It was to be fixed to a **camera obscura**, a blackened box-like apparatus which was used by painters for the reproduction of complex images (the image was inverted on the base and it simply had to be sketched by hand). Initially simple **biconvex lenses** had been used in these dark chambers, but the problem with them was that the images they produced were clear in the middle and blurred around the edges; Wollaston therefore introduced the **meniscus** which was convex on one side and concave on the other, and to which he adapted a **diaphragm** on the concave side. With its equal focal distances the meniscus gave much better definition. Nevertheless, still before the advent of photography, it was also seen to have problems: the very nature of the glass meant that there was a difference between the visible external image and the one which was projected inside the chamber; there was appreciable **lateral chromatic aberration** which produced coloured fringes on the outer parts of the field; the **astigmatic deformation** became a problem beyond a half-field of 20°; the **optical distortion** made straight lines appear curved. Finally, the last fault, which appeared when the Wollaston device was used in photographic instruments, was that the aperture was limited to f/11, due to **spherical aberrations**. In 1821 the Frenchman Charles Chevalier made a **positive lens** using a special glass called **crown-glass**, which has a low **refractive index** and weak **dispersion**. He stuck it on to a negative lens made of lead glass, or **flint-glass**, which has a high refractive index and high dispersion. This was the first **double-lens** objective; it eliminated one of the main faults of the previous one, the chromatic aberrations, but did not eliminate the astigmatism, which remained at the edges. The focal distance also remained limited to f/11.

It is appropriate to pay homage to the Englishman John Dollond, who, in 1757, was the first to tackle the problem of the chromatic aberrations. In doing this he went against the authorities on the subject, authorities which were all the more tyrannical for being personified by the illustrious Sir Isaac Newton. This scientist had in fact identified the problem of chromatism and affirmed in his 1704 *Treatise on optics* that it would be illusory to eliminate the chromatic aberrations by making objectives consisting of two lenses with different refrac-

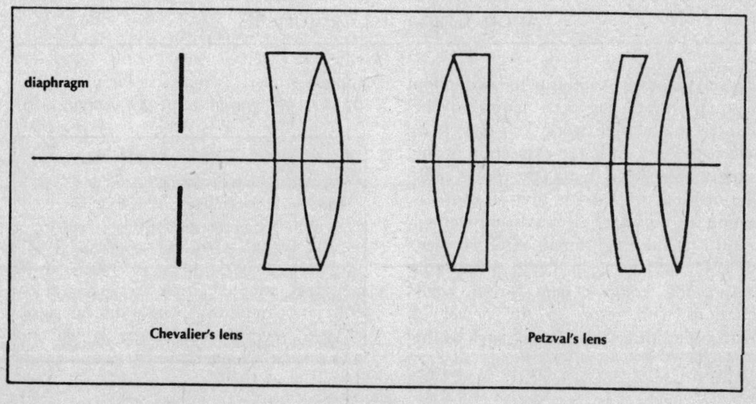

tive indexes, as one might have believed. Despite this postulate, which was obviously false, real achromatic objectives had already been commercialized, in 1733, but, by a common human failing which frequently appears in the history of sciences, no one took any notice of this; all continued to believe, with Newton, that two lenses of different refractive indexes would require an **infinite focal distance**. Dollond himself, who manufactured objectives, defended it against Euler, but in the end he returned to the demonstration made by a Swedish scientist of the falsity of Newton's calculations. The evidence, consisting of existing achromatic objectives as well as the Swedish calculations, eventually encouraged Dollond to make his own achromatic objectives, using one lens of crown and one of flint.

Chevalier's lens had been invented by the time the Austrian portraitist Joseph Petzval resolved to increase the **speed** which had been used until then, and in 1841 he succeeded in using two spaced and modified **telescope lenses**. The two lenses were achromatic and their spherical distortions were noticeably eliminated. Petzval could be claimed to have, in short, added a modified telescope lens to the Chevalier lens (see illustration). The definition around the edges was undoubtedly not good, but it was not a problem inside the picture and was actually almost desirable. The essential point about it was that, at f/3.6, it allowed a previously unknown photographic speed. Even at apertures as large as f/1.6 this objective maintained an excellent definition up to 5° of the axis, which was remarkable.

Rain gauge

Towneley, 1677

The first attempt to measure atmospheric precipitation seems to have been made by the Englishman Richard Towneley in 1677. He installed a funnel on the roof of his house in order to collect the precipitation in a flask by means of pipes. By dividing the amount of water collected by the number of square inches of the base area of the funnel's cone he obtained the pluviometrical rate.

Square roots

Anon, China, 1st century BC

The oldest known example of the extraction of square roots has been found in the Chinese compilation work *Nine chapters on mathematical art*. In it the extraction of the number 1 860 867 is found (the root is 123). The method proposed is that of approximation by increased decimals, reinvented in 1819 by the Englishman W.G. Horner. In 1245, however, in Ch'in Chiu-sao's *Mathematical treatise in nine sections*, some equations with terms to the power of greater than three were found, such as the following:

$$-x^4 + 763\,200\,x^2 - 40\,642\,560\,000 = 0.$$

The first known solutions of **cubic roots** were proposed in China by Wang Hsiao-t'ung (Xiaotong) in the 7th century, six centuries before the Italian Leonardo Fibonacci, who is usually considered to be the first to have found them. It is generally assumed that Fibonacci would have known about them, perhaps via the Arab mathematician who was his teacher while his father was consul in North Africa.

Statistics

Bernoulli, before 1713; Condorcet, 1785 and 1805; Laplace, 1812; Gauss, 1812; Bolzano, 1837; Quételet, 1835; Maxwell, 1860; Fries, 1886

Understood in terms of a technique of mathematical interpretation applied to phenomena of infinite number or complex structure, statistics is derived both from **probability theory** and from the **law of large numbers**. No mathematical discipline has known wider use, for it is used as much in the human sciences, sociology, economics, psychology and politics, as in the pure sciences, astronomy, physics, chemistry and biology; it is as much the basis for **opinion polls** as for **medical epidemiology**.

Probability theory and the law of large numbers were probably invented very early in the history of mathematics, notably in the Mediterranean, for Archimedes set down the bases, for example, of **calculus** in the 3rd century BC. It must be assumed, however, that the double tyranny of Platonic and Aristotelian thought meant that the human brain slowed down for a long time, for it was not until the end of the Renaissance that mathematicians such as Luca di Paciuolo, Gerolamo Cardano and Niccolo Tartaglia began to take an interest in the mathematics of **games of chance**. The bases of probability theory were enlarged by the Frenchmen Blaise Pascal and Pierre de Fermat, and by the Dutchman Christiaan Huygens.

The genesis of statistics contained the germs of an ideological conflict which continued in the 20th century. This conflict was between a determinist conception of nature, which could be said very schematically to go from Laplace to Einstein, and an indeterminist conception which went from Bernoulli to Heisenberg. For Einstein, for example, 'God doesn't play dice', but for Heisenberg the **principle of uncertitude** stated that it was impossible to be sure that a given particle would go from A to B and not from A' to B'. For the 'indeterminists' it was impossible to stand back completely from the subjectivity of the observer.

It was the Swiss Jacques Bernoulli, however, a member of a prodigious line of eight mathematicians, who must be affirmed to have invented probability theory in the modern sense. The theorem found in his *Ars conjectandi*, which was published posthumously in 1713, is stated thus: let us suppose that an event **a** takes place independently of its coincidence with **h**, in other words, the instances of **a** are independent of their probability; let us suppose also that this probability is **p**. It follows that the principles of addition and multiplication can be used to calculate the probability that event **a** will take place **n** times **h** with a relative frequency in the interval **p**±. The most probable value of the frequency of **a** on **n** occasions is that which is the nearest to **p**. After some time, **a** will take place with a relative frequency corresponding to its probability, or, in other words, the probability that the relative frequency will deviate **n** times from its probability **p** less than a given quantity, which approaches the limit of one when **n** is increased indefinitely.

This theorem, called Bernoulli's theorem, was the first of a series of propositions which Simon Denis Poisson would in 1837 call the **laws of large numbers**.

In 1785 Marie Jean Antoine Nicolas Caritat, marquis of Condorcet, was able to further the probability theory in an essay which he improved in 1805 and republished under the title *Eléments du calcul des probabilité des décisions rendues à la pluralité des lois* (Elements of the probability theory and its application to games of chance, lottery and human judgment), which placed the bases of statistics into **politics** and **sociology**.

It was not surprising that the astronomer and mathematician Pierre Simon, marquis of Laplace, subsequently became interested in probability theory. He did so from two different viewpoints: a mathematical one in his *Théorie Analytique des probabilités* (Analytical theory of probabilities), published in

1812, and a philosophical one in his *Essay* on the same subject, published two years later. Being of strictly deterministic conviction, Laplace maintained that a mind capable of knowing all the information about a phenomenon, the famous '**Laplace's demon**', would be able to predict the outcome of this phenomenon exactly. Laplace enriched probability theory and statistical technique considerably by defining them as a **theory of errors**, that is, **deviations from a given average**.

One of the most remarkable mathematicians in an age which had many of them, was Adrian Marie Legendre. Legendre, who was generally unrecognized in his time and unjustly scorned by Laplace, was to refine this approach in his *Théorie des nombres* (Theory of numbers) which was published in two volumes in 1830.

The German Carl Friedrich Gauss, who was heralded posthumously by his peers as the 'prince of mathematicians', advanced probability theory immensely when he was between the ages of fourteen and seventeen. It was at this unlikely age that he demonstrated that restrictions have to be established in order that an infinite series converges towards a determined finite limit. This demonstration was founded on the treatment of the **theorem of the binomial**, where Gauss proved that when **n** is not a whole number greater than zero, one can end up with arithmetical absurdities, such as $-1 =$ infinity. Since Gauss was born in 1777, it was therefore between 1791 and 1794 that he proved his genius. It was in 1812, however, that he showed the possibility of applying the probability theory to the statistics of the phenomena of life.

In 1837 the Czech Bernard Bolzano introduced the notion of **scope** into probability theory. This was based on the **exclusive alternatives of a proposition**, which in a way extended Gauss's theorem.

The Belgian astronomer, mathematician and statistician trained by Laplace, Adolphe Quételet, was one of the founders of the modern science of statistics in that he defined the methods of application as much in the administrative domain as in the sciences. It is owing to him that the method of taking a **census** was learned. As organizer of the first international conference on statistics, he also set down the contemporary notion of '**average man**'. **Criminology** owes to him the treatment of crimes and offences in a statistical manner. Although it is difficult to assign a final date to his work, which occupied him until his death, one of the most outstanding may be cited as 1835, the date on which his work *On Man* appeared.

Statistics then was well advanced, or at least it had come a long way since Bernoulli, but a great many mathematical and philosophical notions still remained to be clarified. Basing their work on Bernoulli's theorem, numerous mathematicians relied in their work on the idea that in the long term events took place at a rate proportional to probabilities. This was a false interpretation of Bernoulli's theorem, however, which postulated simply that it became more and more likely that the frequency would coincide with the probability, independently of the number of probabilities. The mistake proceeded from a deterministic, mechanical and, in short, Laplacian distortion of statistical reasoning. This tended to assume that by dint of measuring the frequencies of a given phenomenon, this frequency would end up being determined with an almost infinite precision, which would have allowed the passage from a subjective conception of the probability to an objective perception. In fact, infinite measurements of the number of bubbles of gas in a given quantity of boiling liquid would provide only an average number of these bubbles. This was the reason Bernoulli had introduced the notion of '**moral certitude**' attached to the statistical calculation where the gap between the relative frequency and the probability was minimal. The Englishman R. Leslie Ellis took the credit for denouncing this illusion in 1843.

Thereafter probability theory, together with the theory of statistics, entered the domain of philosophy, which had been abandoned since the time of Laplace. The probability of the laws in fact became philosophical, given that the number of elements which would have had to be known was unknowable, since it was infinite. Even the gap between the frequency and the probability could not be brought to a minimal fixed value, since these elements were unknown. It was therefore necessary to distinguish between the **natural probability** and **mathematical probability**, as numerous mathematicians did until the 20th century. In doing this they followed the proposition outlined in 1886 by the German Jakob Friedrich Fries. This proposition was based on the work by the Scot James Clerk Maxwell in 1860 on the underlying probabilities of distribution and velocity of the energies of gas molecules.

Stereoscopy

Wheatstone, 1838; Brewster, 1849; Helmholtz, 1851

It is not certain whether the realization of an optical instrument which gives the illusion of relief by the superimposition of two identical images was derived from a discovery or an invention. The fact remains that the first **stereoscope** definitely was an invention, which the Englishman Charles Wheatstone based on the following optical principle (which was perhaps established after the discovery): the superimposition of two images creates an impression of relief provided that the distances separating their centres is the same as the interpupillary distance. The stereoscope made by Wheatstone in 1838 for artistic purposes comprised two mirrors which were inclined so that the two virtual images were superimposed just as in ordinary vision.

In 1849 the Scot David Brewster made fundamental changes to Wheatstone's stereoscope, by having two symmetrical and off-centre lenses enlarge the two images in such a way that it was their real images which were superimposed. The Brewster system was a little more difficult to regulate, given that the lenses had to have a focal distance allowing the images to be observed without straining; not everyone has the same interpupillary distance, which meant that any two users of the stereoscope could obtain two different prismatic powers.

This was the problem that was tackled in 1851 by a great optical specialist of the time, the German Hermann Ludwig Ferdinand von Helmholtz, who made a stereoscope which allowed the gap between the lenses to be altered using a screw. It was Helmholtz who, on discovering that the **luminous spectrum** could be made up from red, green and purple, laid down the foundations for another type of stereoscopy. By the superimposition of a blue-green image and a magenta image, which was obtained directly using bi-colour lenses, it was in fact possible to obtain the illusion of relief from pairs of images which were superimposed and moved forward slightly, the one blue-green and the other magenta. For this there had to be opposing lateralisations to the glasses, for example if the blue-green image was on the left, it was the right-hand lens on the glasses that had to be the same colour. This kind of relief image is called **anaglyph**.

Theodolite

Hipparchus, 2nd century BC; Hero, 1st century BC

The theodolite, a modern instrument which is used for measuring **reduced horizontal angles**, **zenithal distances** and **azimuths**, has its origins in an antique invention which is thought to have been redesigned again within a century. It was in the 2nd century BC that Hipparchus, one of the great astronomers of ancient times, invented the **diopter**. This consisted of a set of two **telescopic lenses in line** which were fixed at the ends to an axle. This was fixed in turn by a pivot to a vertical axle mounted on a tripod, and the whole thing could move horizontally and vertically due to **verniers**. Since it allowed the measurement of **reduced angles**, Hipparchus used it for measuring the **apparent diameters** of the Sun and the Moon. A century later, Hero of Alexandria equipped it with both a **water-level** which enabled the horizontal of a point of observation to be established, and a circular plate divided into degrees which could be fixed at the required angle in any oblique plane. This device was designed for civil engineering and military use, and in particular it enabled the digging of straight tunnels through mountains. It thus prefigured the theodolite.

Hero's modification of the diopter could provide information as to the approximate time when this great inventor lived. In his book *Dioptra* Hero proposed an original way of calculating the distance from Rome to Alexandria, which consisted of observing the same eclipse of the Moon from the two towns and measuring the interval of time between the two observations (this assuming a certain precision in the **clepsydra** — see p. 82). Hero, who was well-informed in astronomy, did not make this suggestion by chance. He must have known that between the year 200 BC and the year 100 BC there would be only one eclipse that could be observed in both Rome and Alexandria, the one which took place on 13 March in the year 62. It would seem likely, therefore, that Hero suggested an experiment in advance which he knew could be carried out, and that he perhaps meant to do so. From this it can be deduced that Hero lived at this time and that it was then that he attained his intellectual maturity, in which case he would have been born in the last years of the 1st century BC and not at the beginning as has previously been supposed.

Thermometer

Ctesibius, 3rd century BC

The first known evidence of this invention dates back to Ctesibius, Hero of Alexandria's teacher: this was the **thermoscope**, an instrument identical to the one which has been wrongly attributed to Galileo. It was a glass U-tube with a vessel at each end, each filled with a liquid; when one of the vessels was heated, the liquid rose because of expansion. This invention could be older, perhaps even by a century, as certain writers attribute it to Philo of Byzantium. The fact remains that a picture of the instrument has been found in a 13th-century manuscript. The use of it must have been somewhat vague, for there do not seem to have been any **graduations** used at the time; the device could have been used to estimate the temperature of the air or that of a sick person. The one reinvented by Galileo in 1592 was not much different in principle apart from the fact that the glass tube was straight and not connected to a vessel, and its free end was submerged in a bowl filled with wine spirits.

The first attempt at graduation was carried out by Galileo's colleague Sanctorius Sanctorius or Sanctorio. In 1611 he fixed a minimum which was determined by the **temperature of melting snow**, and a maximum, which was that of **simmering** over a candle flame. The difference was divided into 110 degrees. The next version of the **thermoscope**, made in 1632 by the Frenchman Jean Rey, differed from the preceding ones only in the inversion of the instrument, with the tube still having only

The thermometers commonly used these days are more or less the same as those which were already in use in the 17th century. They are rather unstable and above a certain temperature their graduations differ slightly from one thermometer to the next. They are nevertheless accurate enough for common needs.

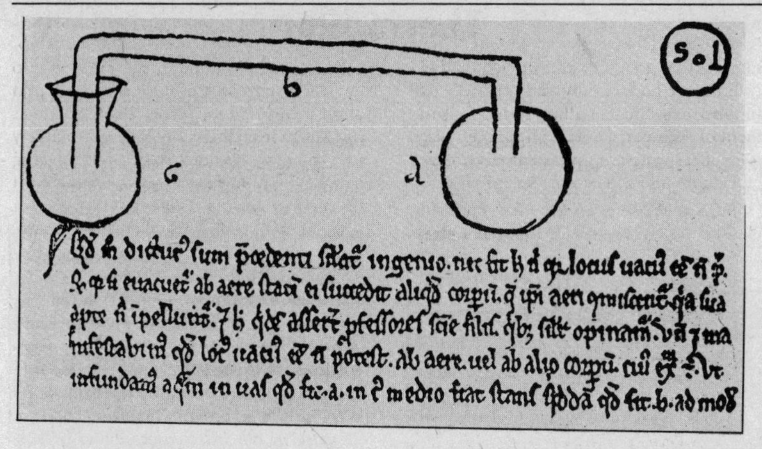

Air thermoscope *of Philo of Byzantium; from a Latin manuscript*

one vessel, full of liquid, and one end free. The thermoscope was no longer an **air thermometer** but a **liquid thermometer**; it bore more similarity to a **barometer**, since it was subject to the effect of **atmospheric pressure**, a problem which was corrected by the grand duke Ferdinand II of Tuscany who sealed the upper end of the tube. From the 1650s onwards the first experiments with the mercury thermometer took place. In 1672 the Frenchman Hubin experimented with a sealed thermometer which had two liquids, **mercury** and **copper nitrate**, with the second liquid eliminating the differences due to changes in the air pressure; in 1695 a thermometer with three liquids and following the same principle was made by the Frenchman Guillaume Amontons.

Boyle, Newton and Huygens endeavoured to improve the thermometers, the graduations of which often did not correspond from one instrument to the next. Moreover, at the time there were more than 30 systems of graduation, which did not make measuring any easier. Around 1700 the Dane Ole Romer began a standardization of the thermometer by fixing the boiling point of water at 60°, and the freezing point for ice at 7.5°. This graduation, which placed blood temperature at 22.5°, was taken up again by the German physicist Daniel Gabriel Fahrenheit between 1700 and 1730. He refined it by fixing freezing point at 32° and body temperature at 96°, with the boiling point of water being around 212°. Fahrenheit subsequently amended the temperature of a healthy person to 98.6°. The precision of his thermometers, which all corresponded to one another, was to popularize both his method and his system of graduation considerably. In 1730 the Frenchman René Antoine Ferchault de Réaumur rationalized Fahrenheit's graduation somewhat by establishing the freezing point of water in an alcohol thermometer at 0°, with the boiling point at 100°. He thus established the first centigrade graduation in thermometry.

This graduation, precise though it was, remained empirical. It was Charles and Louis Joseph Gay-Lussac around 1800 who conferred more scientific bases to thermometry, by using as a foundation the observations of the Englishman Boyle in 1661. A gas at constant pressure increases in volume in a tube according to the rise in temperature. This was the starting point for **gaseous thermometry** but there was still the problem of there being no ideally stable gas. Better precision was attained thanks to the work of the Englishman William Thomson, Lord Kelvin, who calculated the temperature differences on the bases of the **Carnot principle**, that is, referring to the number of joules added to or subtracted from a system.

Thermostat

Drebbel, c.1730

The first mechanism for **automatic temperature regulation** seems to have been invented and made around 1730 by the Dutch mechanic and chemist Cornelius Drebbel. It consisted of a **mercury thermometer** inside which there was a float with a system rods attached to it. The thermometer was first used to control the temperature of a boiler; it was half submerged in the tank and when the temperature of the water in the tank rose, the level of mercury caused the float to rise correspondingly. The system rods then controlled the closing of the damper of the fire. In 1771 the Englishman William Henry invented a **water-column thermostat**, and the **bubble regulator** was invented by the Scot James Watt in 1790 in order to maintain constant pressure in the first steam engine boilers. Watt simply adapted a system which had been used in some mills for a long time.

It was research in alchemy that inspired Drebbel to invent the thermostat. Indeed, according to him a constant temperature had to be maintained for a very long time in the incubators in which he was trying to change lead into gold. The word 'thermostat' was not invented until 1839, by the Scottish chemist Andrew Ure.

Medicine and Health

In general the tendency is to think that medicine progressed continuously and in almost linear fashion from antiquity onwards. This was not the case at all. Ancient medicine was empirical and until the ban on the anatomical dissection of corpses was lifted, virtually nothing was known about the human body. From the Renaissance until the end of the 19th century most of the great advances made in the medical field were related to discoveries, particularly discoveries about anatomy. The best a doctor could offer was a correct diagnosis, and even this was a rare occurrence. Strictly medical inventions were rare, accidental and secondary, therefore, such as the invention of the syringe.

It was in the practice of medicine, hygiene and public health that inventions came to light. For example, Saladin invented the neutrality of military medicines and started up the Red Cross organization. However, time and again it is found that certain principles of cleanliness were practised by ancient civilizations and then forgotten. The quality of 'pharmaceutical products', if the potions and other solutions of olden times can be labelled thus, began to be controlled. The declaration of contagious diseases was made compulsory; in short, the foundations of what was later to become modern medicine are to be found scattered everywhere. It is certain that the Chinese outlined endocrinology, and the Hindus, plastic surgery. In the end, however, it must be admitted that it is essentially only because of vaccines that medicine is in a position to cure and even to prevent disease. Its modesty is laudable in this respect, and only charlatans or cranks promise to heal serious diseases.

This may seem a negative lesson, but it is also positive, for it proves that an invention must be based on knowledge, which is much more important to it than discovery. There is no invention without research.

Ambulance

Larrey, 1792

The idea of the systematic transportation of wounded and sick people in a specially adapted vehicle was conceived in 1792 by the baron Dominique Jean Larrey and put into practice during the invasion of Italy in 1796–1797. At the same time Larrey set up a body of ambulance drivers consisting of **surgeons** and **stretcher-bearers**.

Previously, covered vans, which were badly equipped or hardly equipped at all, had existed for the transportation of the wounded from the battlefield, but neither their number not their design had guaranteed their efficacy.

Larrey's flying ambulances had two or four wheels and were of a unique design. They had a fixed mattress on the floor and padded sides. Each two-wheeled ambulance could take two wounded people, while the four-wheelers could take four. In each ambulance division there were twelve light vehicles, eight two-wheelers and four four-wheelers, and also four cars called the **pansantes**, for carrying the dressings, which were a similar model to that of the arms transportation vans.

Asepsis

Collins, 1829; Holmes, 1843; Semmelweis, 1845; Skoda, 1856 ...

Until well into the 19th century, most doctors and surgeons had absolutely no rules about **hygiene** in the treatment of patients and the handling of wounds. This was due to the lack of knowledge about the existence and particularly the infectious power of **bacteria**. One of the most tragic consequences of this ignorance (which could have been avoided by studying the four hundred or so observations made on bacteria by the Dutchman Leeuwenhoek) was the high mortality rate due to **puerperal fevers**. Authorities such as Lieutaud and Broussais were still explaining it during the 1850s as caused by the 'bad flow of milk' ...

Ignorance about the most elementary rules of hygiene meant that during the wars in the Crimea and in Turkey, from 1854 to 1856, the number of the wounded who died from infection was almost equal to the number killed on the battlefield. Nevertheless the majority of the people in charge continued to scoff at the theories on asepsis.

The real progenitor, if not of asepsis then at least of hygiene, was the Irishman Joseph Clark. During the 1790s in Dublin he had managed to lower the mortality rate due to puerperal fever by maintaining a certain level of cleanliness in the maternity hospitals; his son-in-law Robert Collins pushed these practices even further by having the hospital rooms disinfected using **chlorine**, and then by having the bedding itself disinfected, which aroused the anger or the scepticism of his colleagues. In 1843 the American Oliver Wendell Holmes was the first to understand the infectious nature of the puerperal fever, affirming that even apparently clean hands could transmit the infection. The idea was labelled as ridiculous. In 1845 the Hungarian Philippe Ignace Semmelweis, a pupil of Skoda and Rokitansky, important professors of the time, realized that the mortality rate was very different at the hospitals of the obstetricians Bartsch and Klin: with Bartsch, where the delivery was carried out by **midwives**, there were fewer deaths than with Klin, where it was carried out by students. After an inquiry he concluded that the mortality rate was higher with Klin because the students did not wash their hands beforehand. He installed wash-basins at the doors of the obstetric clinic and impressed upon the students the need to wash their hands. These precautions seemed monstrous and Semmelweis was dismissed a few months later. Driven mad by the persecution, he died in 1865. In 1849, however, his patron Skoda, who was impugnable, took up his defence, together with another professor Hébra. The enemies of **asepsis** were still not disarmed, however.

Asepsis did not compel recognition until after it had proved its virtues, or at least those of **antisepsis**, which were established in 1862 and published in 1867 by the Englishman Joseph Lister who advocated cleaning wounds with **carbolic acid**. Pasteur's work during the 1870s on the role of microbes would gradually defeat the final, stubborn adversaries of asepsis.

Autopsy

Galen, 2nd century; Vésale, after 1537; Harvey, 1628

However unpleasant it might be, the **anatomical dissection** of corpses, or **autopsy**, is the basis of anatomical and physiological knowledge about the human body; before it these areas were dominated only by confused and muddled ideas. For the duration of the Roman Empire, however, the dissection of human bodies was forbidden by law. The first doctor to realize that autopsy was necessary was the famous Greek Galen from Pergamon, the pupil of the contemporary teacher Pelops, who went to live in Rome in 146. Galen was doctor to the gladiators, and would have attained the essence of his knowledge and discoveries from observing his patients' wounds, as well as from the dissection of monkeys and pigs. He might also have had the opportunity to study some decomposed corpses, but it is more likely that Galen carried out secret dissections of human corpses, probably protected by his own prestige and the benevolence of the emperor Marcus Aurelius and then of his son Commodus. Some of his observations, on the **laryngeal nerve** for example, add considerable weight to this supposition.

Some thirteen centuries after Galen, anatomy had hardly progressed at all; this was because of lack of dissections, since in this respect Islam was of no more help than Rome. It is said that the emperor Frederick II had two men sacrificed after a meal so that they might be autopsied, but this anecdote is rather doubtful, when one considers how refined this prince of Hohenstaufen really was. He had founded the University of Naples and the School of Medicine at Salerno, and Italian poetry was born and the arts flourished at his court, as noted by Dante. Around the 15th century, nevertheless, autopsy was authorized more and more often in the case of **suspicious deaths**. For example when the pope Alexander V died suddenly in 1410, an autopsy on his body was ordered. Autopsy had in fact lost some of its prohibitions since the publication of a manual on anatomy by the Bolognese Mondino in 1316. Despite this it had a long way to go to become standard practice, and in 1536 the famous Flemish doctor André Vésale, who was studying at the University of Louvain, was reduced to running campaigns to try to find bodies of

Nowadays autopsy is hardly used at all for the enrichment of anatomy or physiology, but it is still indispensable for **legal medicine** and remains fundamental to **medical epidemiology**. For example, samples taken in autopsies revealed that the bacteria which cause **Legionnaire's disease** existed at least ten years before the epidemic which attracted international attention to this pulmonary infection, and that the **AIDS** virus also existed ten years before the epidemic of this disease broke out in 1980.

criminals. When he became a recognized doctor in 1537 he created a chair of anatomy and surgery, this time setting up autopsy as a **fundamental medical discipline**. His prestige enabled him to obtain the necessary corpses from the burgomaster of the town. This great doctor can be said to have been the founder of autopsy — and he would have been able to practise at his leisure at the court of Charles V, where there were many deaths due to syphilis (when not to poison).

One of Vésale's pupils, the Englishman William Harvey, initiated the progression of autopsy from the stage of anatomical study to that of **physiological study**, as attested by his *Anatomical study of the movement of the heart and the blood* which was published in 1628. By basing his work on the Italian Realdo Colombo's propositions and by combining **clinical study** with anatomical examination, Harvey was able to decipher on outline of the circulation of the blood.

Contraception

Fallopio, c.1550; Wilde, 1823; Mensiga, 1880

For many centuries various forms of contraception (other than the **coitus interruptus**) were used by several peoples, particularly the Egyptians, the New World Indians and the Chinese. The most common method consisted of impregnating **vaginal tampons** with supposedly spermicidal substances such as vinegar, or with substances which had real anti-ovulatory power like the **Lithospermum ruderale** used by the Mexican Indians and rediscovered in 1941 by the American Marker. It was only around 1550 that the Italian Gabriele Fallopio, professor of anatomy at Padua University, thought of using washed pig intestines as **condoms** to prevent the spread of syphilis. This invention seems to have been quite successful as a means of contraception.

The next invention in this domain was that of the **diaphragm** by the German F.A. Wilde in 1823; he effected this by using a mould of the neck of the uterus to make hoods out of metal and rubber crêpe. This idea was taken up again in 1880 by his compatriot W.P.J. Mensiga, who made the first diaphragms relating to the modern definition of the term, which were larger and fully cast from **vulcanized rubber**. It is important to note that in the 2nd century the Greek gynaecologist Soranus of Ephesus, who worked in Rome, had made a distinction between contraceptives and **abortifacients**, which proves that both of them were already in use at the time. It is also interesting to note that neither were used very much, for they were expensive, inconvenient and sometimes dangerous. This was the reason for infanticide being so widespread.

The only product of traditional pharmacopoeia to have proved its contraceptive virtues is the root of *Lithospermum ruderale*. This was used during the 1950s as a basis for the American Gregor Pincus's research and it resulted in the making of the first **contraceptive pill**.

Dentistry

Neferites, c. 2600 BC

The origins of dentistry are much older than one might think, for dentures have been found on Egyptian mummies; these were either gold or silver **ligaments** which kept in place **artificial teeth** made of ivory, or gold **crowns** which were intended to strengthen a loose tooth by using a healthy tooth to support it. The name of the first known maker of false teeth is Neferites, who lived around 2600 BC. The Egyptian and Mesopotamian 'dentists' also made use of **gimlets** to pierce abscesses. Furthermore, traces of tooth fillings using various substances such as **terebinth resin** have been found on these mummies, but it is unlikely that it was known how to clean a decayed tooth at that time other than by scraping with a metal instrument.

The first known manual of dentistry was published in Leipzig in 1530, and it was not until 1580 that dentistry courses opened at the University of France in Paris. The title of dental surgeon appeared at the beginning of the 17th century, but the prestige it carried does not seem to have amounted to much. Modern dentistry can be thought to have been born at the time of the treatise published in 1728 by Pierre Fauchard which was the first to lay down the foundations of a specific discipline. It is interesting to note that until about 1750 women were allowed to practise as dental surgeons in France, but this authorization was then taken away from them.

Toothpastes and toothbrushes have existed since ancient times; the latter were made of couch grass and the former were made up according to an infinite number of recipes, by mixing **abrasives** such as chalk or powdered charcoal with juices, pastes, crushed herbs, etc, up to and including human urine, the ammoniac content of which was used by the Romans in the 1st century BC and was rediscovered by modern toothpaste manufacturers.

The first dental plates probably appeared in Germany in the 16th century and were made of boxwood. The less painful dental plate made out of vulcanized rubber was invented by the American Charles Goodyear in 1840 and it came into use in 1854.

Doctor's neutrality

Saladin, c. 1187

After the Hippocratic oath, the second section of the ethical code of a doctor to be internationally acquired was his political neutrality. This was established by the sultan Saladin in Jerusalem, after his victory in the crusades in 1187. The Ayyubid potentate actually stipulated that the doctors on the battlefield should come to the aid of all the wounded, whether they were Muslims or Christians. He even organized his own doctors to visit the camps of prisoners and, for the first time, authorized foreign doctors to circulate freely in the enemy camp. It was in this way that Richard the Lionheart's doctor, Ranulphe Besace, was authorized to visit the wounded Christian prisoners. This initiative anticipated the founding of the **Red Cross** by the Swiss Henri Dunant by seven centuries.

Electrical cardiac stimulation

Anon, Greece, c.5th century BC; Nysten, 1802; Steiner, 1871; MacWilliam, 1889

The idea of the electrical stimulation of the cardiac muscle, which was effected in the second half of the 20th century with the **pacemaker**, must be very old. From the 5th century BC the Greeks applied torpedo fish, or **gymnotes**, on the thorax of sick people in order to stimulate vital reflexes in them. One can only treat the results of such operations reservedly, for they were much more likely to hasten death than to delay it. The fact remains that well before the electrical nature of nervous transmission was understood, and before the coordination centre of the cardiac cavity contractions (**Keith and Flack knot**) was discovered, when the useful **intensities** and **rhythms** of cardiac stimulation were not known and no distinction was made between **continuous** and **alternating current**, doctors launched themselves into the beginnings of cardiac stimulation. It was in 1802 that the Frenchman Pierre Henri Nysten obtained — with some difficulty — some freshly guillotined corpses in order to succeed where Xavier Bichat had failed before him: to restart a dead person's heart. He was the first to observe that the **auricles** could move after the other parts of the heart had ceased to do so. He was undeniably the father of electrical cardiac stimulation, although there is the account of an English doctor in 1774 who 'resuscitated' a child which had fallen from a second floor and was apparently dead on the pavement by administering electrical charges on the chest. It is an interesting story but the hypothesis behind it is all the more doubtful for the fact that at the time it was still not really known how to distinguish between a coma and death.

Having succeeded in restarting the hearts of 55 different species of animal, in 1871 the German F. Steiner dared to apply his technique to a patient anaesthetized with chloroform who had fainted and on whom other methods of revival had failed; he transmitted a weak electric current through a needle which had been implanted directly in the heart and, surprisingly enough, he succeeded.

New progress was made in 1889 thanks to the American J. Mac William who recognized the need for a **rhythmical stimulation with an alternating current** and the necessity to apply this current to a very precise place in the heart. Mac William carried out an operation without drawing blood by applying **electrodes** to the thorax and he also succeeded in restarting a heart which had stopped. Thereafter much progress ensued and, two years before the end of the century, the Frenchman Auguste Chauveau was the first to introduce a probe or **catheter** into the carotid artery of a living horse, which he pushed against the flow of blood up to the left ventricle of the animal and consequently recorded a cardiac stimulation.

Electrotherapy

Scribonius Largus, c.1st century; Dioscorides, 2nd century; Galen, 2nd century; Paul of Aegina, 7th century; d'Arsonval, 1891

The use of electricity for therapeutic purposes preceded both the understanding and the mastery of electrical phenomena by a long way. It was around the 1st century that the doctor Scribonius Largus, who practised in Rome, mentioned the advantage of shocks given by **torpedo fish** in the treatment of chronic diseases. His assertions were taken up in several slightly different ways by his colleagues Dioscorides in the

2nd century, Galen, also in the 2nd century, and Paul of Aegina in the 7th century. The shocks from the torpedo fish could be powerful and repeated, both in the water and in the open air, and they were stronger in tepid water than in cold water. Of course we cannot know what effect they could have had on illnesses such as dermatosis, which had not yet been identified, gout, headache or melancholy. Therefore it is difficult to use the term 'invention' precisely regarding the first attempts at electrotherapy. The fact remains that these doctors must have noted some beneficial effects which are extremely difficult to comment on now.

In 1891 the brilliant French electrical engineer Arsène d'Arsonval undertook to apply **high-frequency currents** to patients who had 'nutritional diseases', diabetes, gout, rheumatism, obesity, etc. This therapy was to be pursued by d'Arsonval and his followers until well into the 20th century.

Today it is not possible to discuss the effects of '**d'Arsonvalization**', as it was called. The reason for this is firstly because none of the illnesses listed (which also included cardiovascular and respiratory troubles) are what d'Arsonval thought them to be, diabetes possibly having several origins, including genetic and viral ones, rheumatism possibly being an auto-immune disease, gout being a metabolic illness, etc. Secondly, the actual descriptions of the effects of d'Arsonalization were vague as well.

The fact nevertheless remains that electric currents, either directly or by means of the **magnetic fields** that they generate, do have an effect on living organisms.

Of the possible effects of electricity, only one has been verified medically: that the healing of broken bones will be accelerated by currents of very low frequency and intensity. The effect of **electric shock** on disturbed psychic states is very much contested. The possible effects of the **magnetic fields** generated by high-voltage cables, which have been studied since the 1970s, seem to be real, but whether they are harmful is controversial.

Endocrinology

Anon, China, 2nd century BC

It seems to have been in the 2nd century BC that the Chinese, obviously without the use of any fundamental medical information, thought of extracting certain substances from human urine for therapeutic purposes. Urine in fact contains **steroid hormones**, oestrogens and androgens, and **pituitary hormones**, the gonadotrophins, which stimulate the sexual system. These hormones were extracted in crystalline form by reduction, and the crystals were called '**autumn ore**'. A recipe dating from 1025 describes this extraction process in detail, using quite considerable quantities of urine (600 l) which underwent evaporation and then sublimation. The Chinese managed to

A rather intriguing detail is that in order to precipitate the hormones contained in the urine, the Chinese used relatively modern methods, such as precipitation by **calcium sulphate** or by **natural saponins**, a technique which was rediscovered only in 1909 by the German Adolf Windaus. It is also worth noting that numerous primitive pharmacies prescribed urine for therapeutic purposes until not very long ago. In 1930 in Nubia and Upper Egypt it was recommended that **baldness** be treated with poultices of a pregnant woman's urine, following a very ancient recipe, whereas it was only during the 1960s that it was established that baldness could be caused by an excess of **androsterone**.

obtain in this way several grams of crystalline hormone extracts, free of urea. These extracts were incorporated into pills which were for the use of women who were apparently sterile, among others. It was only in 1927 that Ascheim and Zondek discovered that the urine of pregnant women is in fact rich in steroid hormones. It could therefore be said that the first, albeit empirical, experimenters in endocrinology were the Chinese. In this case they drew their inspiration from alchemical principles.

Endoscope

Segalas, 1827; Helmholtz, 1851; Killian, 1897

The first medical implement designed and made for the study of internal cavities of the body seems to have been the **speculum**. This was a small mirror fixed to the end of a rod which was illuminated from a distance by a candle and from which the reflection then illuminated the cavities being studied, in this case the larynx and the female urethra. It was invented by the Frenchman Pierre Segalas in 1827. In 1851 the German Hermann von Helmholtz invented a spyglass mounted on an inclined mirror which focused the light from a lamp placed near the patient on to the back of the patient's eye, thus enabling the ophthalmologist to examine the eye through the spyglass (a device which was later replaced by a concave mirror fixed to the ophthalmologist's brow which had a hole in it through which he examined the back of the eye); this was the first **ophthalmoscope**. Edison's success in making very small electric light bulbs in 1878 enabled more sophisticated instruments to be made, such as the **electric cystoscope** by the German Max Nitze that same year, and then the German Gustav Killian's **bronchoscope**, a rigid tube which was pushed down the trachea to the bronchial cavity.

Forceps

Chamberlen, c.1630; Levret, 1747; Smellie, 1751; Tarnier, around 1870

The **obstetrical forceps** designed to make childbirth easier in certain situations are generally thought to have been invented around 1630 by Peter Chamberlen, the founder of a famous English family of obstetricians. The original nature of the invention is debatable, since **pliers** had existed since ancient times: for gynaecological examinations doctors used pliers called **retractors**, the design of which varied considerably from one practitioner to the next, since the instruments were individually made to order. The Chamberlen forceps were straight with blades, designed to grip the head of the foetus during delivery. This developed into the hook called the **head-pull**, the insertion of which was more gentle. The secret behind it was jealously guarded until around 1710, when Chamberlen's bankrupt son, in exile in the Netherlands, sold it to a colleague named Van Roonhuysen; it was actually very simple, since it consisted of a locking lever between the two blades, which enabled traction to be exerted on the head of the foetus without the pressure on it being increased. Van Roonhuysen was less protective of his secret than Chamberlen, and, at least in principle, the forceps began to be known throughout the world of obstetrics. In 1747 the Dauphine's obstetrician André Levret changed the straight blades of the forceps into **spoons** and bent them in such a way that they could be applied to the foetus in the **upper birth canal**. In 1751 the Englishman William Smellie modified the curvature allowing an even greater flexibility in use. Henceforth the models of forceps increased in number, with each obstetrician making his own following his own design. The main improvement was made around 1870 by the obstetrician Stéphane Tarnier. He invented forceps with hollowed-out spoons which had a special part perpendicular to one arm of the instrument, the **tractor**, which allowed traction to be exerted in a single direction even more easily than was possible with Chamberlen's locking forceps.

Hyperthermia

Anon, Egypt, 2500 BC; anon, France, around 1495

The treatment of diseases by the artificial elevation of the body temperature is a discipline which reappeared in France during the decade of the 1970s for the treatment of **tumours**, particularly deep-seated tumours, under rigorously controlled conditions. An Egyptian papyrus dating back to the 3rd millennium mentions the application of hot plates for the treatment of tumours. This method seems to have had various setbacks in the frequency of its use until the 20th century, but it certainly featured at the end of the 15th century in the repertory of **syphilis** treatments. The patient with 'Naples sickness' was placed in a sort of iron cabin which was heated using a fire. Evidently it was in an empirical way, on experimental bases, that hyperthermia was practised long ago, for the **role of fever** in activating the **immune system** was not known at that time.

Military medicine

Julius Caesar, 48 BC

Julius Caesar was the first to have the idea, in 48 BC, of appointing a body of 17 doctors to each legion of 7000 men, assigning these doctors the rank of non-commissioned officer so that they were dependent upon the public treasury; thus he created **army medical officers**. The most numerous, therefore, were the legion doctors, referred to by the name and the number of the legion. Then followed the cohort doctors who treated the praetorian, urban and look-out cohorts. The camp doctors, divided into nurses and surgeons, were assigned the treatment of the garrison soldiers. Rome had also created auxiliary army medical officers who acted as nurses. The most famous of the first army medical officers was Scribonius Largus who accompanied the emperor Claudius on his military mission to Great Britain.

Orphanage

Maguebodus, around 581

Since both adultery and abortion were considered to be crimes in the first centuries of Christian Europe, there were many abandoned newborn babies. In 442 the council of Vaison insured against this desertion by recommending the '**exposure**' of the child by a third party for the purposes of adoption. The recommendation was undoubtedly effective, for the council of Mâcon began to use it in 581, and the most famous, if not the first, '**orphanotroph**' or orphanage was opened shortly afterwards by St Maguebodus.

Pharmaceutical control

Al Mu'tasem, c.1250

Since an apothecary's products were precious, they soon became the objects of fraud, such as the replacement of opium with wild lettuce which contained a much weaker sedative. It would have been around 1250 in Baghdad that the botanist and chemist Ibn el Baïtar was given the previously unknown position of **general apothecary inspector** (mouhtaseb), with inspectors under his command. This was instigated by the last caliph of Baghdad, Al Mu'tasem.

Repair surgery

Susruta, 49 BC

The oldest available proof of **repair surgery** dates back to the Hindu surgeon Susruta who in 49 BC carried out an operation which was rather audacious when one takes into account the knowledge of the time. It consisted of treating **intestinal perforations** and **obstructions** by joining together the damaged or occluded part of the intestine after cutting into the abdomen below the navel. Susruta sutured the segments in what may seem to be rather a strange way: he placed the freshly cut heads of giant black ants on the edges of the opposing sections. This process is all the more surprising for the 20th-century observer because it shows a knowledge of the powerful antiseptic properties of **formic acid** (H. COOH) secreted by the ant heads; these properties were particularly useful in the treatment of highly septic wounds.

The same Susruta was also the first to carry out a **rhinoplasty** operation. He used a technique which consequently became classic, of the **lowering of a triangular frontal flap**. After this heals into the nasal flesh, which takes a few days, the nasal peduncle is cut.

Sterilization by steam

Anon, China, 10th century

It seems to have been in the 7th century BC that the Chinese thought of **fumigating** their houses using the smoke from certain plants to get rid of the insects; this practice was subsequently to be extended to manuscripts, for killing the worms in them. Perhaps this should be seen as the invention of **sterilization** by steam, as recommended by Lu Tsan-ning (Lu Zanning) in 980 in his *Discourse on the investigation of things*. This author stated specifically that the belongings of people who had died from epidemic fevers should be immediately and entirely treated with steam in order to stop the disease spreading to the family. This practice is all the more surprising considering that **microbes** and their role were totally unknown at the time. The Chinese therefore seem to have sensed the existence of the elements of contamination and the destructive action of **high temperatures** on these elements.

Stethoscope

Laennec, 1816

Even though the act of listening to the noises of internal organs, particularly those of breathing, seems very old, as it dates back to Hippocrates in the 5th century BC, the invention of an instrument to amplify it dates only from the 19th century. It was in 1816 that the Frenchman René Théophile Hyacinthe Laennec invented a wooden tube which both isolated the organic noises from the surrounding environment and made them louder. They could, however, only be listened to with one ear.

Around 1839 the Czech Joseph Skoda improved the stethoscope by equipping the instrument with two hollow tubes which meant that both ears could be used.

The stethoscope was called a '**plectoriloc**' at first and it came in two kinds, both made by Laennec: a hollow tube for the examination of the thorax, and a solid tube for the examination of the heart. The first quickly developed into two models: one was cylindrical, and used for listening to changes in the voice; the other was widened at the auricular end, and used for listening to breathing and rattles in the throat. When it was presented to the Academy of Medicine in 1819, the instrument was greeted with so much scepticism that in his *Medical Science Dictionary* which appeared around 1820, Mérat classified it as a useless invention. The first person to defend it, strangely enough, was the writer Chateaubriand.

Syringe

Pascal, 1646; Pravaz, c.1835

It was during his research on fluids that Blaise Pascal, in 1646, designed and made a syringe for scientific, not medical, use. This syringe seems to have been the inspiration for the administration of purgative **clysters**, but it was still a long way from the hypodermic syringe. The hypodermic syringe was derived from observations on aneurysms made around 1830 by the French doctor Velpeau. Velpeau had in fact noticed that sticking a needle into an aneurysm could cause the formation of a clot, which constituted the basis of one method of treating this venous malformation. Then came the idea of introducing chemical substances which would accentuate the effect of such injections, and it was at this time, for the purpose of injecting ferric chloride, that the French doctor Charles Gabriel Pravaz invented the **hypodermic syringe**. It consisted of a silver cylinder, and the lid of the pump body could be screwed down; the cylinder had a nozzle to which the cannula could be fixed, which was itself extended by a steel trocar. Each half turn of the screw pushed out 1/30 of a gram of liquid. The **glass syringe** was invented in 1895 by a glass-blower called Fournier. In the meantime the syringe had been improved: instead of the liquid being ejected by turning a screw, there was a **piston** which could be pushed down much more easily; also, the needles had become interchangeable.

Pravaz, an orthopaedic specialist by profession, was a daring innovator, and so he carried out the treatment of various illnesses, such as asthma and tuberculosis, by injecting compressed air using a device which he had invented himself.

Vaccination

Anon, China, 10th century BC

The anti-smallpox vaccination was rediscovered in the West in the 18th century in the form of '**variolation**' but it actually seems to have been invented in China by a Taoist monk, whose name has not survived, in the 10th century BC. It was probably invented following philosophical principles and was simply a case of treating an evil with an evil. Thus an archaic form of variolation was practised some thirty centuries ago, and it is worth mentioning for it already included the principle of attenuation of the germs by multiple transmissions. Vaccination took place not by **scarification**, but by the implantation of a **pad** carrying some attenuated germs in one nostril.

According to Chinese texts, the cultures of attenuated germs were kept in hermetically sealed flasks and placed away from heat and light, which indicates a degree of biological knowledge which was astonishing for the time.

Transport

The evolution of transport since humanity began is dominated by one peculiar fact: that mechanical energy, available since at least the 1st century BC in the form of Hero of Alexandria's aeolipile, the prodigious steam engine prototype, did not begin to be studied in a practical way until the end of the 18th century.

Nothing was lacking for the invention of boilers and locomotives in the time of the Roman Empire, or during the Byzantine Empire. Hard metals, pistons, cylinders, connecting-rods and bit-braces all existed. Even the universal joint had been invented. However, the upheavals which followed the barbarian invasions and the fall of Rome, the intellectual obscurantism which prevailed until the end of the Middle Ages (engineers permanently ran the risk of lawsuits for sorcery), and then the incessant wars which ravaged Europe, the most destructive of which was the Thirty Years' War, all left technology and science in a pitiable state. To invent was a luxury of the solitary person, and was generally without much future; it was only in the 18th century that intellectual curiosity really came alive again. Then the first steam-powered vehicle of modern times was built. This was the famous *fardier* which was in fact no more than an artillery tractor, built by an artillery officer, Nicolas Joseph Cugnot. Fourteen years later the era of aerial transport was inaugurated at Versailles by a sheep, a cockerel and a duck on board a hot-air balloon which was built by some stationers, the Montgolfier brothers. Humankind had only to await the end of the 19th century for the internal combustion engine to arrive.

The golden age of steam was splendid, but short-lived: it lasted hardly a century, from the mid-19th to the mid-20th century. It did however have the huge advantage of assuring the expansion of industrial mechanization and of allowing rapid transport.

Asphalting of roads

Anon, Mesopotamia, 1st millennium BC; anon, France, c.1730

Before anyone knew how to build **roads** (see p. 203), there was obviously a great deal of mess caused by **rain-water** sinking into the ground and many problems with **dust** during dry weather. One slow and costly solution was **paving** but this was reserved only for the main roads. In the Middle East, beginning with Mesopotamia in the 1st millennium BC, wherever there were **oil-bearing rocks**, **bitumen** or **asphalt** were used as an impermeable covering. The use of bitumen for this purpose seems to have stopped at some unspecified time, although it was still used for the **caulking** of ships, as indicated by the techniques for the extraction of bitumen which were described in the 16th century by the German Georg Bauer, also known by his Latin name Georgius Agricola. It was only in the 18th century that asphalting of roads began to spread again, invented or reinvented by an unknown French engineer. Presumably it was the abundance of bitumen obtained when the first oil-field was exploited near Pechelbronn in France, beginning in 1745, that rekindled the idea of using it. Even when the roads had been built, the asphalting prevented the water from seeping down into them and meant that the road would last longer.

Balloon and airship

Montgolfier brothers, 1783; Giffard, 1851

The first ever balloon flight was made at Annonay in 1783 by the brothers Etienne and Joseph Montgolfier. They had been struck by Priestley's ideas on gas, and had also been the first to have the idea of filling a hollow geometrical structure with air.

The following year, the Frenchman Guyton de Morveau tried to control the

The success of the Montgolfier brothers' experiment does not necessarily mean that they are to be attributed with the invention of the balloon. At the beginning of the 15th century the German Konrad Kyeser, one of the great inventors of the end of the Middle Ages, had designed a fantastic craft which really was a **lighter-than-air flying machine**; it was a wooden frame of somewhat reduced size, since a rider could carry it at the end of a lance, and it had paper hanging from it. A **powder burner** generated hot air which lifted the craft, which was in the shape of a dragon. This peculiar aerostat was in fact an incendiary device, the mouth of which carried a flask with a burning fuse and a mixture made up of one part oil, four of sulphur and one of bitumen. Obviously we cannot judge whether it was effective, but nevertheless it was conceived on the principle of **hot-air ballooning**.

The origin of this principle, which undoubtedly predated Kyeser, is unknown; in any case, the principle was widely known at the time of the Montgolfiers. Indeed, a few years before them the Brasilian priest Joao Gusmao had also tried to make a lighter-than-air craft, but in vain.

The **first international aerial voyage** of all time was accomplished in 1785 by the Frenchman Jean-Pierre Blanchard, one of the pioneers of the **parachute** (see p. 202), and the American John Jeffries. Their craft was a balloon with a rudder powered by four aerial paddles. They arrived at the end of the Dover-Calais crossing almost naked, having had to throw out their clothes to unballast ... This trip came two years after the **first human flight**, which was made above Paris by Francois Pilâtre de Rozier and the marquis of Arlandes, and it preceded by eight years the flight made by Blanchard above Philadelphia, which was the **first American flight**.

direction of a balloon with the help of a **rudder**; this inventor can therefore be considered as the direct forerunner of the Frenchman Henry Giffard who in 1851 was the first to design and make a **lighter-than-air** craft. This was equipped with a system of **mechanical propulsion** by steam, a **propeller** and a rudder: Giffard had invented the **airship**.

Bicycle

Théson, 1645; Sivrac, 1690; Blanchard and Masurier, 1771; Drais de Sauerbron, 1818; MacMillan, 1839; Michaux, 1855–1861; Starley, 1874; Lawson, 1874

The first ever vehicle which was not pulled by horses but was powered by human energy was a wooden **quadricycle**, for which the Frenchman Jean Théson obtained a 30-year patent in 1645. This enabled him to improve the vehicle, which was moved only by the action of pushing on the ground with alternate feet. It was very heavy and in 1690 it was reduced to two wheels (by Sivrac). The machine, the body of an imaginary animal, was still rather inconvenient, however, as was the following version of it which was made in 1771 by Francois Masurier and Michel Blanchard: they had gone back to Théson's four wheels, which were definitely more stable. The **two-wheeler** reappeared in 1818 with the baron Drais de Sauerbron's **dandy horse**.

On 31 May, 1868 the first cycling competition in the world took place in the Saint-Cloud Park. It gathered together two hundred cyclists and was won by the Englishman James Moore who cycled 135.207 km in 10 h 25 min, an average speed of 12.960 km/h ... and that was on a bicycle with solid tyres!

The real technical innovation in this domain came from the Scot Kirkpatrick MacMillan who, in 1839, invented a system of **swinging pedal-cranks** placed on either side of the front wheel which transmitted a movement to the back wheel by means of two light **connecting-rods**. In this respect the real inventor of the bicycle should therefore be MacMillan, despite the fact that he was not very successful. The problem with his invention was that the **pedals** did not rotate, but followed circular arcs and were tiring for the knee joints. The invention which really advanced the bicycle was that of the Frenchmen Henry and Pierre Michaux in 1861; they made the cranks of the pedals pivot integrally around the axle of the front wheel, to which the pedals transmitted their movement directly. This was the first **velocipede**, of which the Michaux brothers manufactured 142 models (and which was quickly copied). In 1874 the Englishman James Starley invented the **spoked wheels** and the **metal frame** which made the machine appreciably lighter. In the same year his compatriot H.J. Lawson invented the transmission to the rear wheel by an **endless chain**. **Gear-wheels** were to come later.

Brakes

Anon, England, 1767–1776; Westinghouse, 1869; Smith, 1872

The name is not known of the engineer who designed the first recorded braking system, around 1767, when the first cast-iron rails had been installed and the speed of coal wagons increased. It consisted of **brake shoes controlled by rods** which were applied laterally on the rim or vertically on the hooping of the wheel. This principle has remained fundamentally unchanged to this day; only the **tightening system** has been improved. The first brakes used on motor vehicles were those on locomotives and, in the first trains, which went at low speeds and carried relatively small loads, the braking of the locomotive was enough to slow down the wagons as well. It remained thus until 1870.

At this time the braking was designed according to the following general mechanism: two shoes, which were hollowed out according to the arc of the wheel which they had to stop, were set on a pair of opposing and articulated levers. These were controlled in the centre by a raising lever and the whole thing was fixed to the chassis. This lever was then connected to the driving position by two articulated shafts; the lowering of the transverse shaft, which was caused by the corresponding lowering of the vertical shaft, using a wheel, lifted the central lever, thus causing the blocking of the wheels. The brakes were hand-operated, therefore, which firstly required a certain amount of manual strength, and secondly, a flawless coordination between the operation of the brakes and the railway signals. Methods of mechanical braking were sought until 1869, when the Englishman George Westinghouse invented **compressed air brakes**, with which the pressure exerted by the pistons of a cylinder of compressed air could be applied in return for much less effort. In 1872 the Englishman Francis Pettit Smith introduced the system of braking by **pneumatic pressure** which was assisted by the power of the locomotive.

After the locomotive, the first vehicle to have brakes fitted to it was the bicycle.

Diving suit

Anon, Greece, 3rd century BC; Kyeser, c.1400; Borelli, 1679; Lethbridge, 1715; Klingert, 1797; Siebe, 1819 and 1830

The invention of the diving suit seems to be relatively old, for as early as the 3rd century BC Aristotle had mentioned devices which enabled divers to stay under water for a long time and breathe normally; the philosopher described metal containers which appear to have been the helmets of primitive diving suits. These had two holes in them into which glass lenses were inserted and they had to be kept upside-down, adjusted on the shoulders, to stop the air from getting inside. in order to enable the divers to remain for a 'long time' under the water and therefore to breathe, they had to be supplied with fresh air from the surface, which could only be done by keeping the mouth at a tube which emerged at the surface, or by pumping this air through a tube. As rubber did not exist then this necessarily flexible tube was presumably made out of canvas waterproofed with bitumen, for example, or out of leather, Aristotle does not say.

It seems reasonably certain, moreover, that Alexander the Great himself went down underwater in a device called a **colimpha** about which we can only conjecture. The most plausible possibility is that it was a large **diving bell**, the distant ancestor of the **bathyscaphe**, with a paned-glass opening on top, for the colimpha is described as allowing light to enter; it would have descended vertically in order to conserve the air trapped inside. In any case, neither the diving suit described by Aristotle, nor the bell used by Alexander could allow a descent of more than 3 m, if only because of the pressures exerted at this depth or greater. In Aristotle's diving suit these would have required pumps of a power which did not seem to be available at the time. Moreover, the vertical thrust exerted on the bubble of air would have prevented the **independent divers** from going down, unless they were ballasted. The same reservations apply to a 4th-century drawing by the Latin author Vegecius which shows divers equipped with a helmet and breathing through a leather tube attached to a bladder on the surface, which acted as a float. This drawing too is only plausible provided that the tube was in the diver's mouth, with the aqualung only allowing him to keep his eyes dry. If this were not so, as with Aristotle's suit, the law of connecting vessels would have meant that the diving suit and the tube would both have become filled with water.

This type of diving suit seems to have survived until the Middle Ages, for the English philosopher Roger Bacon reported in 1240 that there were 'instruments which men could use to walk on the bed of the sea or of rivers without endangering themselves'. Around 1400 the German Konrad Kyeser, who also cites Vegecius, showed two types of diving suits in a drawing of two divers facing each other: one, which he obviously borrowed from Vegecius, consists of a helmet made of an unknown material and enabled the diver to breathe through a tube attached to a bladder on the surface; the other, which was much more original, was covered with a large helmet which had two glass holes and was fitted to a **close-fitting tabard** fixed to a belt. This equipment, which heralds the modern diving suit, is astonishing because it is the first known representation of the body suit; its usefulness can only have been to protect the diver's body from very cold water. However the air supply was limited by the available power of the pumps, assuming, it must be said, that it was air pumps that were used for the supply.

The next major step in the development of the diving suit seems to have been taken in 1679 by the Italian G.A. Borelli, who described for the first time a suit equipped with a supply of air in a leather **bottle** which was carried on the back, and a **piston** for regulating the **specific gravity**. The invention attests to a real intuitive genius; nevertheless one cannot help but be sceptical about its reliability.

Commercial genius was displayed by the Englishman John Lethbridge, who in 1715 took up the idea of Kyeser's suit almost exactly as it was, in his own name, and made a fortune from it. It hardly benefited aqualung technique, however, any more than did the invention by the German K.H. Klingert who in 1797 combined Lethbridge's suit and Borelli's **dorsal air tank**.

The diving suit entered the modern age with the apparatus designed in 1819 by the German Augustus Siebe: this consisted of a **metal helmet** fitted to a **leather suit**, and, of particular interest, it was supplied with air by a **pressure pump**. The air injected kept the water-level below the diver's chin, provided that he remained more or less vertical. As it was, this suit enabled underwater work to be carried out. It was improved by Siebe himself in 1830: this time the diving suit was complete and watertight, and the helmet had an opening for **air to escape**. This invention rendered incalculable services in the building and repairing of maritime works and structures, and Siebe continued to make improvements to it until his death in 1872. The Siebe suit could be used on a large scale and up to depths in the region of a hundred metres owing to a pumping system invented by the Frenchmen Rouquayrol and Denayrouze.

It was only in the second half of the 19th century, notably owing to the research by the Frenchman Paul Bert explained in *La Pression barométrique* (The barometric pressure) in 1878, that it began to be understood that diving with a simple air supply below 5 m presented dangers which increased with the depth. The divers who went down much further than this suffered from nervous troubles, sometimes fatal disorders, which were caused firstly by the forced injection of **carbon dioxide** into the body tissues under the effect of the pressure, and then, during decompression, by the formation of bubbles of **nitrogen** in the tissues. It was not until the 20th century that the need for decompression by stages, and then the need for **specific gas mixtures** during the dive, were understood.

Harness

Anon, China, c.2nd century; anon, Europe, around 1150

The oldest animal harness dates from China in the 2nd century and was a **breast harness**; it included a part which was attached to the animal's chest and was probably designed for oxen. There may however have been harnesses before then. The **shoulder collar** appeared later, at an unspecified date, and did not reach Europe until the 10th century, whereas the breast harness did not become widespread there until the 12th century, at about the same time as the **swingletree**. It would seem that the first harnesses, dating from the 4th to the 1st millennium BC were the yoke and collar harnesses, but these must have been significantly more painful for the animal and therefore less effective.

Helicopter

Anon, China, c.500; Vinci, 1480; Borelli, 1680; Paucton, 1768; Launoy and Bienvenue, 1784; Ponton d'Amécourt, 1862; Forlanini, 1877

Although no machine propelled by a **horizontal propeller** managed to take off before the 20th century, the invention was in gestation for a long time. Some 5th-century Chinese toys with horizontal propellers fell at a slower speed than the speed with which they had been thrown into the air. A famous drawing by Leonardo da

'Helicopter' *Drawing by Leonardo da Vinci (1452–1519)*

Vinci dating from 1480 shows a flying machine to which a kind of large **Archimedean screw** turning on a vertical axis had been fixed. In 1680 the Italian Giovanni Alfonso Borelli seems to have made a model which allowed him to improve the theory of lifting by a horizontal propeller, as did the Frenchman Alexis Jean Pierre Paucton, who in 1768 returned to studying the problem with a model with two propellers. The absence of a motor, however, rendered this research purely academic. In 1784 Launoy and Bienvenue made a flying toy which was a model of a helicopter without an engine. The arrival of the steam engine meant that the Frenchman Gustave de Ponton d'Amécourt and the Italian Carlo Forlanini (the former a numismatist like Paucton, and the latter a doctor like Borelli) could, in 1877, make reduced size models which were the precursors of piloted helicopters.

Hull with watertight compartments

Anon, China, 2nd century

The division of the hull of a ship into watertight compartments, which did not appear in the West until the 18th century, was invented in China in the 2nd century. Chalk and coal-tar were used to **caulk** the compartments. The number of compartments increased according to the dimensions of the boat (there could be more than 24 partitions) and also depended on whether the boat was for sea or soft-water navigation. The oldest evidence proving the great age of this invention, which was essential to the security of the boats, was the 20 t junk dating from the Tang dynasty (7th–10th century) which was discovered in 1973 in the province of Jiangsu.

Hydraulic piston motor

Anon, China, 530

One of the most disconcerting inventions in technological history is the **hydraulic piston motor**, which appeared in China at the beginning of the 6th century. It consisted of a **monocylindrical motor** with a **piston** activated by a rod linked to a **camshaft**, which itself was powered by a **mill-wheel** by means of a **driving belt**. A description of it appears in a work dating from 530 on the Buddhist temples and monasteries in the province of Luoyang. The invention probably predated this therefore, although its exact date is not known. It was used as a **bellows** in a smelting works. The first graphic representation of it appeared in 1313 in Wang Chen's *Treatise on Agriculture*. The invention is disconcerting because it foreshadowed the **steam engine** without using steam and it was almost identical to a similar type of motor which was reinvented in 1757, twelve centuries later, by the Englishman John Wilkinson and which was adapted for steam by his compatriot James Pickard in 1780. Moreover, the Chinese hydraulic motor displays a sizable technological breakthrough in its introduction of the **transformation of a circular movement into a rectilinear movement** which was touched upon by the mechanics of the Alexandrian School.

It is thought that the re-inventions of the Chinese piston motor derive from descriptions of Chinese machines transmitted by various intermediaries, such as the Italian Agostino Ramelli (1588), as well as from travellers' descriptions and original Chinese documents.

Lighthouse

Anon, Greece (?), 1st millennium BC; Ptolemy II, Philadelphus (?), 285 BC; Louis de Foix, 1584; Argand, 1782; Fresnel, 1820; Smeaton and Douglas, 1882

The installation of a light on a coastal hill to act as a **landmark** for navigators is an invention which has progressed through the ages. It originally consisted of simple fires which were lit on hills and kept going, and since Homer mentioned them in *The Iliad* and *The Odyssey* in the 9th or 8th century BC it can be deduced that the custom had already existed for some time. The first building designed to shelter a light seems to have been the famous **Alexandrian pharos**, one of the Seven Wonders of the World, the building of which was undertaken in 285 under the orders of the Egyptian king Ptolemy II Philadelphus. It was about 115 m high (it was destroyed by an earthquake). The Romans built lighthouses systematically to act as landmarks for their navigators, and in the 4th century there were about thirty of them, from the Atlantic to the Black Sea. The light source was provided by a **torch**. The fall of the Empire put a stop to the building and even the upkeep of lighthouses. It was only in the 12th century that they were erected again. In 1584 the engineer and architect Louis de Foix undertook to rebuild the lighthouse erected by Edward, Prince of Aquitaine. The gradual collapse of the island of Cordouan into the estuary of the Gironde made it into the first **open-sea lighthouse** when the building was finished in 1611. It was a magnificent monument, 35 m high.

The usefulness of the first lighthouses was proportional to the power of their light source; in other words, for a long time it was very modest and somewhat shaky, with the visibility of the relatively meagre light given out by two dozen candles across the sea being reduced still more by the sea-spray. The Swiss Aimé Argand must therefore be credited with improving the usefulness of the lighthouses significantly by his use of an oil lamp with a **circular wick**. This was fanned by a central inflow of air and had a metal chimney above it to provide a draught. The strangeness of the town councils concerned meant that Paris, where Argand had hoped to be given the privilege of making this type of lamp, rejected this remarkable invention (which was copied in another form by the pharmacist Bernard Quinquet), and that the disappointed Argand (the privilege had been granted him but to no avail) went to offer his invention to England where its benefit to the illumination of lighthouses was recognized. Argand's invention had a future, for it was completed by Francois Arago and Augustin Fresnel, first of all by simply increasing the number of circular wicks.

In 1763 the English amplified the light source by using **parabolic reflectors**, many-sided mirrors made of silvered glass, around the flames, and in 1820 Augustin Fresnel finally made a huge improvement in the lighthouses by inventing the **graded lens**. An outstanding optician, Fresnel in fact used an idea first thought of by the famous Georges Louis Leclerc, count of Buffon. A wide-angle lens capable of absorbing the whole field of light from one source would be so thick that it would absorb too much light, whereas a lens shaped as though it were made up of a superimposition of concave lenses would be much lighter whilst having a considerable diffusion power. In actual fact the hollowing-out of the lens was to lead to the creation of **circular prisms** surrounding the central plane surface.

The advent of **electricity** from 1850 brought with it a satisfactory solution to the problem of the intensity of the light source. Thereafter the problem of the shape of the towers was tackled. This required research into two areas of which very little was known at the time, the first being that of **swell**, which was just one aspect of the **force of the sea**, the second being that of **wind**, the force of which had also been misunderstood.

The research into swell prompted a

revision of the shape of the towers and the way they were built. The composition of two undulatory wave movements at consecutive moments comprises a difference of one tenth, which means that the highest ones are every tenth wave and consist of two movements joined together. This results in the sea exerting a double thrust every tenth wave, attaining a height which is double that of an average wave. Furthermore, the resistance provided by the tower or lighthouse could mean that the wave reached double the height of the building. The lighthouse would be surrounded by a watery curtain which would not only dim its light but also threaten its structure. A basic shape had therefore to be established such that the tenth wave would break without reaching the tower, and the structure of the tower itself had to be strengthened according to the residual forces which were just beginning to be recognized. In 1882 the Englishmen John Smeaton and James Douglas cleared the way for the rational construction of lighthouses. These two engineers had learnt from their own previous difficulties and in particular from the collapsing of several lighthouses.

It was Léonor Fresnel who, at the end of the 19th century, demonstrated that the **pressure of the wind** at 200 km/h reached 275 kg per m^3 on a cylindrical tower and that for such a building to be able to resist the joint pressure of the wind and the waves it was necessary to design structures capable of resisting five times the pressure. This would be 1375 kg per m^3, a decisive calculation, but one which underestimated the real force exerted by some storms.

Locomotive

Trevithick, 1804; Stephenson, 1813 and 1829; Seguin, 1828; Crampton, 1840

The invention of the locomotive constitutes a special chapter in the history of the steam engine and motorized vehicles, for it threw up new and specific problems for the engineers. At the beginning of the 19th century the value of the **steam engine** was no longer in doubt, for it had been proved by the first successes in marine propulsion. Moreover, **railway tracks** already existed: they were used especially for the transportation of coal, drawn by animals (in 1720 about twenty thousand horses were used in this way for pulling goods wagons), and there had been no problem with their adoption, for haulage on an even surface was much easier than on roads. It seemed that the steam engine simply had to be adapted to pull the wagons on the railway tracks.

In fact no one knew how to build steam engines which had a small enough **weight** and enough **power** to pull the wagons, and the problem of how to keep them on the rails had not been solved either.

The ancestor of all locomotives is undoubtedly the one which was built in 1804 by the Englishman Robert Trevithick to travel a distance of 9 miles (13 km), and which weighed 5 t. His experiments were satisfactory, for it managed to pull a 25 t load at 5 mph, (a little less than 8 km/h). However, it rested on cast-iron rails which tended to break. The vehicle was astonishingly archaic and could have been built in the 1st century (see illustration). Its very long **connecting-rod** was particularly picturesque.

Trevithick built other locomotives, one of which, weighing 8 t, reached 12 mph or

The first train to run on the Stockton–Darlington line was drawn by George Stephenson's locomotive *Active*, later renamed *Locomotive No 1*. It consisted of the engine, a tender, 6 goods wagons, the director's coach, 6 passenger coaches and 14 wagonloads of workmen.

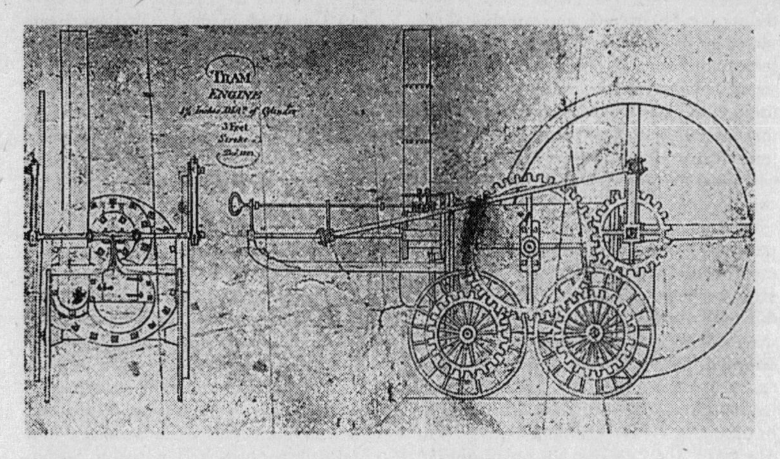

The *first locomotive built by Trevithick in 1803. Contemporary drawing by Llewellyn*

The Rocket*, locomotive built in 1829 by George and Robert Stephenson*

19 km/h, but strangely enough very few people were interested in the project. Until around 1830, and therefore even when the concept of the locomotive had developed seriously, there were still many who supported the horse. Trevithick therefore lost interest in locomotives. In 1811 his compatriot John Blenkinsop took up the initiative again and devised a complicated system, in which a **cogwheel** rested on **racks** fixed to the rails. This system was adopted and was in use for several years without any problem. It was an invention with little future, for it required rails with racks which were both more expensive and more complicated than smooth tracks. However the

distances to be covered then were short, and no one in the 1810s really thought that the locomotives had much of a future ahead of them. Other engineers, such as the Englishman William Hedley in 1814 with his '*Puffing Billy*', tackled the problem and failed for the same reason as Trevithick: the cast-iron rails broke under the weight of the engines.

In 1813 the Englishman George Stephenson, having studied his predecessors' experiments and particularly those by Hedley, built a machine which hardly differed from the preceding machines except that it had **wheels** which were **coupled**, in twos, by **connecting-rods** and it was much lighter, weighing slightly less than 4 t. It did not go much faster, with its 6 km/h or so, but it could pull up to 30 t. Moreover, the rails did not break during the experiments. Its speed was not a determining factor, since it was not really considered important at the time; what caught the manufacturers' attention was the fact that Stephenson's machine, a model which he was to name *Blucher*, had more possibilities than animal traction. In 1825 the **first public railway** in the world was created, the 12-mile (about 18 km) Stockton to Darlington railway.

Stephenson made endless improvements to his locomotive, but the first characteristic improvement to the railway concept was made by the Englishman Timothy Hackworth. This consisted of arranging the cylinders in such a way that the connecting rod was shorter; in addition the **heating surface** of the boiler was enlarged so that the output was increased. Stephenson's *Lancashire Witch* made use of these innovations and in 1828 it reached about 12 km/h pulling 50 t on a 1/440 slope. It was however the *Rocket* which in 1829 imposed the concept of the locomotive as a means of haulage on to public opinion; weighing 4.25 t it won a competition against four competitors by pulling three times its own weight at an average speed of 22.77 km/h on a track of 39.76 km. The principle which enabled Stephenson to win was not his own, however.

This principle had been patented one year previously by the Frenchman Marc Seguin; it was that of the **tubular boiler** which was to become widespread for more than a century. It consisted of increasing the heating surface by making the heat circulate in horizontal tubes arranged longitudinally in the boiler; moreover, Seguin had increased the draught by using a powerful **ventilator**, which Stephenson replaced by a system of injecting the exhaust steam at the base of the chimney. Seguin had built his locomotive before the *Rocket*, but it was not used until later.

The most notable invention which marked the next step in the development of the locomotive was the increase in diameter of the driving wheels which the Englishman T.R. Crampton installed behind the boiler in 1840. This principle was not adopted in England and Crampton built his machine in France. The **Westinghouse brakes** (see p. 193) were finally to ensure the security of railway transport. The locomotive was not changed fundamentally until the **electrification** of the railways in the 20th century.

Metal hull

Anon, England, 1777; Brunel, 1843

Even though the armour-plating of wooden hulls with metal plates is relatively old (see p. 213), the idea of building the ship hulls entirely out of metal did not seem attractive to naval architects until the second half of the 19th century. The theories about **water displacement** were badly defined and the common conception was that metal hulls could not float. Moreover, technological information circulated poorly, and the naval architects who tackled the building of the first totally metal hull would have learned a few lessons in the process; this ship was a passenger launch — with sails — built in 1777 in Yorkshire by an engineer whose name is unknown. Ten years later, also in

England, a 21 m-long barge was launched which was made of wood plated with sheet-metal. In 1821 the first entirely metal steamship, the *Aaron Manby*, made the link between London and Le Havre. It was only in 1843, however, that the Englishman Isambard Kingdom Brunel partially resolved the inherent problems with the metal hulls and **propeller propulsion** by launching the transatlantic liner *Great Britain* which was 98.1 m long, 15.4 m wide and had a capacity of 3689 t. It had four steam engines each generating 1000 hp, coupled to a propeller with six 4.9 m-diameter blades which turned at 53 rpm.

The success of the *Great Britain* partly anticipated the advent of **battleships**, but the supporters of wooden hulls still considered it to be an oddity, a hybrid with its three auxiliary sailing masts.

As well as the objections of the wooden-hull supporters, the first metal hulls had some inherent problems. The first was their **weight**, which meant that they were hardly suited to commercial transport and not at all to military missions; consequently the first battleships had to reduce their armament considerably in order to compensate for the excess weight of either the armour-plating or the metal hulls. The second problem was that they could only be propelled at **speeds** which were compatible with military missions by the use of steam engines, which increased their load even more, both by their own weight and that of the fuel stores. Even for commercial crossings the sailing ships were doubly competitive, in terms of both the speed and the loads carried. A third reason was the **mediocre output of steam engines**. A fourth reason was the fact that the metal hulls deteriorated very quickly and therefore required frequent and costly **refitting** every eight months; nobody had actually thought of painting them yet. There were additional inconveniences, such as the **drag of the propeller** on cargo-passenger ships, which decreased the speed when the ship was sailing, and the **fire risks** linked to the fuel stores.

Owing to their small live load capacity, the first metal hulls tended to gigantic proportions, since this was the only way of competing with the wooden hulls in the transportation of cargoes. This was why, in 1858, Brunel had a boat built which was much bigger than the *Great Britain*, the memorable *Great Eastern*, which was 210.9 m long, an extraordinary size for the time, and displaced a record-breaking 28 145 t. This ship, which was decidedly hybrid for it was powered by wheels, propeller and sails all at once, could carry four thousand passengers, a remarkable number for the time. It was doomed to commercial failure but it nonetheless made its mark in the history of naval architecture, for it was the first to tackle and resolve such technical problems as the pressure of the flow of water against the rudder blade and the first to have a hull designed not according to empiricism and tradition but to the new principles of **hydrodynamics**. The *Great Eastern*, picturesque monster that it was, inaugurated the age of the **transatlantic liner**.

Motorized vehicles

Cugnot, 1765–1770

The first vehicle to be powered by mechanical energy was the Frenchman Nicolas Joseph Cugnot's '**fardier**', a steam-powered tricycle which in 1769 ran for about 20 minutes at a speed of 3.6 km/h. The machine was in fact an artillery chassis where instead of the horse Cugnot had placed a driving wheel activated by a **steam boiler**. This vehicle, a second version of which is in the Museum of Industrial Arts and Crafts in Paris, was really not very practical, since it had to stop every twenty minutes, for a period of twenty minutes, to enable the steam to reach an effective pressure, for Cugnot had not established adequate energy parameters. Although it was abandoned following the disgrace of Choiseul, the minister who financed it, this project was taken up again during the ensuing years and in 1790 a **steam omnibus** was built in Amiens which worked in a much more satisfactory way. There were steam omnibus lines in Paris in 1800. Between 1790 and 1800 the Americans Nathan Read, from Salem, and Apollo Kingsley, from Hartford, also built several steam omnibuses.

The German Otto von Guericke should also be mentioned, for in the 17th century he built a steam-powered vehicle of an astonishingly technical nature which included **cylinders with pistons** and **connecting-rods**, but it has not survived.

One of the most peculiar ancestors of mechanical locomotion was certainly the horse-less coupling which was demonstrated in Paris in 1748 by the mechanic Vaucanson, an automaton specialist. The vehicle's wheels were activated by a system of **springs** and the reduction systems by **cogwheels**. It was set in motion manually, and evidently did not go very far.

Paddle-wheel boat

Zu Chong Zhi, 494–497

The paddle-wheel boat seems to be organically linked to the invention of the **steam engine**. Historical research carried out in the 20th century reveals, however, that this type of propulsion was designed and applied successfully by the Chinese engineer and mathematician Zu Chong Zhi, between 494 and 497, and that it had lasting success into the beginning of the 20th century. There is only one previous known reference to this type of propulsion, dating back to the 4th century; the Latin text which includes it is rather controversial however.

The Chinese paddle-wheel boat, of which several hundred models were made, was obviously man-powered. It was named *Quian Li Chuan* and could cover 1000 li or 500 km in a day, travelling on rivers. It is mentioned in historical annals, for example with regard to the general Xu Shih Pu's campaign against the rebel Hou Jin. Under the Song dynasty (1127–1279) it was equipped with four wheels each with eight paddles. Some of the boats were very large in size, between 60 and 100 m, and could carry seven or eight hundred men, and at the same time some were built which had an odd wheel at the stern. Put to the test on a few transatlantic boats after the advent of steam in the 19th century (and combined with sails as a precaution), the paddle-wheel system was only really popular in modern times on the large navigable waterways such as the Mississippi. Its vulnerability in storms caused it to be abandoned.

Parachute

Anon, China, 2nd century BC

Although the invention of the parachute is commonly attributed to Leonardo da Vinci, it is described in the *Annals* of the Chinese historian Sseu-ma Ts'ien (Sima Qian), which date from 90 BC. This work describes the parachute as being relatively old, and therefore it is thought to date from the preceding century. Joseph Needham, an English specialist in Chinese history, found the description of a jump carried out by a thief from the top of a minaret in Canton using '**two umbrellas without shafts**' in a work dating from 1214.

The first Western reference to the parachute was made by Simon de La Loubère, French ambassador to the king of Siam, in

1687–1688; the diplomat reported that a Siamese was ennobled by the monarch for having jumped from great heights several times using a double device such as the one described above. It is thought to have been this information that gave J.S. Lenormand the idea of repeating the feat, which he did successfully during the 1780s. It was Lenormand who invented the word '**parachute**'. In 1785 Jean-Pierre Blanchard had captivated the public with his release of animals attached to parachutes.

Paved street

Anon, Assyria, 3200 BC; Trésaguet, 1761; McAdam and Telford, 1738–1818; Favier, 1841; Polonceau, 1844

The history of inventions in the domain of road paving, which is crucial to civilization, is most disconcerting. Indeed, whereas thirty centuries before Christ the eastern empires used roads which could withstand heavy vehicles and which meant that traffic could move quickly and reliably, travelling in Europe in the 18th century was extremely hazardous in numerous areas simply because of the state of the roads. They were riddled with potholes and the vehicles on them were always in danger of tipping over or breaking their axles.

The Romans are generally cited as the inventors of modern roads, and it is true that they built the first logical network of **paved roads**, motivated by the extension of the Empire. In the 2nd century BC the *via Appia* linked Rome to Brindisi, for the transportation of the troops from Africa and the orient; about a century later the *via Egnatia* went from Durazzo, which is now Durrës, to Albania as far as Salonica, and in the 1st century the construction of the *via Domitia* began — this linked Rome to Spain as far as Cadiz, going via France. Two great transalpine roads led to Lyons, from which there were four others, leading to Bordeaux, the Channel, the Baltic and the Mediterranean. In Britain a system of roads linked channel ports with other important centres such as London, Chester and York. The modern A1 is based on the Roman Ermine Street and Dere Street. These were paved roads, which were sometimes crossed by channels and built on piles to cross marshy areas. The Romans improved the building techniques a great deal.

The first inventors of the paved road, however, were the Assyrians who are known to have built roads in 2000 BC which had two paved strips 1.5 m apart on which their carts and chariots could run. The discovery in 1928 of an Assyrian chariot dating back to 3200 BC suggests that paved roads were actually built more than a millennium previously. The tradition of doing so was certainly not lost, for in 1500 BC there were paved roads in Crete as well.

The Romans carried the technique of building roads to an unprecedented height; they were undoubtedly the first to understand the need to put down a **hard surface**, in this case slabs of lava or granite. They understood that this should be laid not directly on to the ground, which would quickly become uneven because of drainage, caving-in, and differences in the distribution of pressure, but on to a layer of

> The oldest road in the world which is still in use is the Royal Road of the Incas. It covers the 3600 km from Quito (Equador) to Cuzco (Peru). It was built in the 15th century and in fact consists of two roads, one coastal and the other mountainous, 7.5 m wide. It has numerous tunnels and sections of stonework, including, interestingly, some asphalted parts.

> The development of the railways put a stop to the study of road construction, which did not begin again until the beginning of the 20th century. It was the advent of the automobile in the first half of this century that was to stimulate this technique as much as it had ever been since the first Assyrian paved roads.

gravel which was sufficiently thick to ensure the stability of the surface layer. The Romans invented the **concrete road** too: the 'Iron road' discovered in the UK consisted of slabs of black concrete placed on iron slag and an iron framework. In marshy areas, still in the UK, the Roman roads were built on foundations of **wooden stakes** placed diagonally in order to ensure an even distribution of pressure, and covered with moss in order to guarantee a certain cushioning of the same pressures. After the fall of the Empire not one road was built: for centuries the Europeans lived on the Roman road stock which they were satisfied to maintain, as well as they could.

France was the first country to take up the initiative of the Roman builders, firstly with Jean-Baptiste Colbert, who instituted the corvée on the roads, then with Anne Robert Jacques Turgot, who abolished this same corvée. The central 'Administration of Bridges and Roads' was created in 1716, and the 'School of Bridges and Roads' was founded in 1747. There was still a long way to go. The new roads made at the beginning of the 18th century were in fact layers of sand and stones which were neither calibrated nor drained, terribly cambered and not stabilized. **Lateral drainage channels**, **anchorage posts** and **lateral rubble stones** had not yet been rediscovered; the last of these would stop the road, which had not been covered, spreading. The Romans' fundamental technique seemed to have been lost. It was the head engineer of Bridges and Roads, Pierre Trésaguet, who in 1761 put the road-making technique into practice again; he rediscovered the principle of the two layers used by the Romans, but inverted it so that the foundation layer was made of rubble stones and the covering was chippings.

Then in Britain between 1783 and 1818 John Loudon McAdam and Thomas Telford advanced the technique. McAdam invented the smooth surface which meant the wheels met with less resistance; he did this by putting down paving which was stabilized under the weight of the vehicles. An important point which shows the technical ignorance of the time is that McAdam's ideas were vigorously opposed and the inventor carried out his first attempts at making roads with three layers of stones at his own expense. As for Telford, he stressed the importance of the need for a **foundation**, arguing that in rainy weather the underlying layer of soil turned into mud which flowed up to the surface between the stones; Telford therefore, quite unwittingly, reinvented the Roman technique. The foundation consisted of a 30 cm-thick layer of crushed stone covered with a layer of fine gravel on top of which the surface paving was placed. It was Telford too who fixed the limits of the inclines: 3% for upward slopes, 2.5% for downward slopes.

In 1841 the Frenchman Favier was the first to think of using topographical plotting for the study of road layout, which led to the techniques for the calculation of earthmoving and embankments. Antoine Rémy Polonceau's invention of the steamroller in 1844 brought about great progress in road-making technique as it meant that the layers could be packed down at the same time as they were built.

Propeller

Archimedes, 3rd century BC; anon, China, c.500; Leonardo da Vinci, 1480; Du Quet, 1723; Ericsson, 1837

Few inventions have such confused origins as the propeller, which nevertheless became an essential element in the development of **mechanical propulsion** from the 18th century onwards. According to some writers, its origin was Hero of Alexandria's **screw elevator** (see *Reduction system*) which was inspired by the Archimedes screw; the original idea of the screw would go back therefore to a time between the 3rd and the 1st century BC. This was not the actual propeller, however, since it had no blades.

The first real propeller seems to have appeared in China around the 5th century, on a toy which prefigured the **helicopter** (see p. 195) where it effectively had a propellent function. It is impossible to know if this prototype contributed to the invention of **windmills**, which seem to have appeared two centuries later in Syria and the inventor of which, like the inventor of the Chinese helicopter toy, is not known. The windmill may nevertheless be considered as the first application of the thrust of a fluid to a mobile helicoidal structure, and it was the only example of this until the beginning of the 19th century.

It is possible that in describing the helicopter in its prefigurative version in 1480, Leonardo da Vinci may have invented the propeller; he should therefore be cited as the putative inventor of this device. The first person to have studied the propeller systematically from the mechanical point of view was the Frenchman Du Quet in his paper presented to the Academy of Sciences in 1723. From then onwards several scientists became interested in the propeller, such as the Swiss Daniel Bernoulli in 1753, and the Frenchman Alexis Jean Pierre Paucton, one of the pioneers of the helicopter, in 1768. The first to use the propeller for the propulsion of a vehicle was the American David Bushnell, who fitted an arm propeller for the propulsion of his submarine *The Turtle* in 1776. Thereafter the propeller was used for propulsion in many different ways, which until the invention of the aeroplane were all in the maritime domain: by the Englishman John Fict in 1796; by the American Robert Fulton in 1797; by the Frenchman Frédéric Sauvage in 1832; and by the Englishman Francis Smith in 1839, to mention but a few.

Despite this the propeller was still considered as a marginal oddity, because it could be activated only by a **steam engine**, which had a mediocre output, was not very reliable and needed bulky stores. The marines at every level of the naval hierarchy continued in their preference for sailing ships. Moreover, since **iron hulls** (see p. 200) were not yet widespread, the propellers installed on the wooden hulls caused a serious problem, which was the deformation of the hulls by the appearance of the **fore-and-aft arc**, which threatened to break the **line of the shaft**. The Englishman Isambard Kingdom Brunel was probably the first to consider the metal hull to be a requirement for the successful use of the propellor; in 1843 he fitted a propeller to the *Great Britain*'s metal hull.

But the engineer who caused the propeller to become widespread in naval architecture was the Swede John Ericsson, (who later became a naturalized American) who patented a propeller in 1837. There was certainly no shortage of propeller inventors (in 1867 the Englishman John Bourne counted no less than 470 of them!), but Ericsson, who in 1837 had also built a ship with a propeller in English shipyards, the *Francis B. Ogden*, was the first to prove that this type of propulsion was reliable: his boat crossed the Atlantic without mishap. It is worth noting that this was a boat with a metal hull.

Despite all this the propeller was still not

Half a century, if not more, had to go by after the advent of the propeller for the optimal design to be established. Frédéric Sauvage, one of the first engineers to have pleaded the cause of the propeller as a means of propulsion, was also one of those who advocated the design which had the weakest output: that of a **helicoid with a single whorl**, similar to a section of an endless screw (1832). At the same time the Frenchman Augustin Normand and the Englishman H.T. Barnes, who specialized in the physics of fluids, had made a **propeller with several blades** Comparative experiments served to prove the unquestionable superiority of this latter model. In 1838 the Englishmen Smith and Rennie, who did not find the experiments by Normand and Barnes very convincing, tried to return to the helicoid and tested a propeller of this type which had several whorls, then stated that the effect was better with a single whorl, before agreeing with the evidence of Normand and Barnes supporting a propeller with blades. Smith was in fact helped by chance, for when one of his propellers with a double whorl broke, the ship gained speed being propelled by only one whorl ... It is now known that the best output is achieved with a **large diameter** in proportion to a **low acceleration**.

adopted: until 1861 many shipbuilders were still keen on wheel propulsion, for in that year, in fact, the famous Cunard Line launched its transatlantic liner *Scotia* which had paddle-wheels and sails. This was to be the last of its type, however, for from 1850 the Inman Line, the Norddeutscher Lloyd, the Hamburg Amerika Linie and the General Transatlantic Company launched steamships with propellers, although they still had sails as a precaution. It was only in 1858 that the military marines, singularly lagging behind, finally and unanimously adopted metal hulls, steam-power and the propeller (see *Armour-plating of battleships*).

The first time the propeller was used for aerial propulsion was by the Frenchman Henry Giffard in 1851. He designed and made a steam-powered airship with a propeller and rudder (see *Balloon and airship*).

Rails

Anon, Europe, 16th century; Reynolds, 1768

The invention of rails came before that of **locomotives**; some archaeologists suggest that the regular grooves which can be seen along street gutters, in Pompeii for example, could have been the very first stage of rail development. These would have been to stop the chariots veering off course, to allow more uniform traction so that less animal energy would be required, and consequently to allow greater speed. All the same, the first proper rails appeared in the 16th century; they were designed to reduce the **road resistance** to vehicles which were mainly loaded with coal and were hauled by animals from the mine galleries to the distribution points, or even directly to where the coal was to be used. They appeared more or less simultaneously in France, Germany and England and were made of oak or pine, mounted on sleepers which were also made of wood. The increase in the traffic, due to the increased use of coal, meant that the wooden rails became worn down quickly and so bands of iron were fixed on to the upper surfaces. It was not until 1738, in England, that the first **cast-iron** rails were made. The idea behind them is attributed to Abraham Darby, who had had the idea of melting down the coke ores and who therefore lowered the cost price of cast-iron. In 1768, in order to make the wagons more stable, the Englishman William Reynolds thought of making rails with rims. The shape of the rails changed several times, from the flat rails of 1800 to the present double-mushroom rails, including the inverted-U rails of the first Bordeaux-Sète line.

From the 1850s to the 1920s the rail gauge varied considerably from one country to the next, for both military and economic reasons, forbidding the continuity of numerous national railway networks; in the UK the gauge was 2.139 m, in France 1.439 m, in Spain 1.674 m and in Russia 1.520 m.

Rudder

Anon, China, 1st century

The oldest documents about navigation indicate that ships were steered by two free oars with large blades on either side of the prow, which were turned by a navigator depending on the direction to be followed. These oars were generally fixed at the stern. The rudder fixed to the prow did not appear in the West until the 13th century. It had however been invented in China in the 1st century, if not before, as indicated in a description in the *Shi Ming* or Liu Xhi's *Dictionary of Terms* which appeared in the year 100. Under the Song dynasty (1127–1279) the rudder was modified: it stuck out under the prow in the same proportions as it did at the stern, and sometimes two auxiliary rudders were attached to it. Furthermore, its depth could be regulated by raising or lowering the axle, depending on whether the vessel was in shallow or deep water.

Steamship

Hulls, 1736; Perrier, 1775; Jouffroy d'Abbans, 1783; Rumsey, 1787; Fulton, 1803 and 1807

It was a long and tortuous task to adapt mechanical energy, in this case the steam engine, to the propulsion of boats. There are reasons for this: firstly, the first steam engines were mediocre, being not very reliable and rather low-powered; secondly, sailing ships had ardent supporters, especially in the military domain (see p. 213). Their objections were based mainly on habit and partly on technical arguments, such as the bulkiness of the steam engines and the fuel stores. It was only in the second half of the 19th century that steam was definitively imposed in the navy — after it had been accepted for use on commercial vessels — as a result particularly of the advent of **metal hulls** (see p. 200) and of the **propeller** (see p. 203).

Paradoxically, the first attempts to use steam in the propulsion of ships preceded the **steam machines** by several years, including the fixed machines (for it was obvious that a sea-going engine imposed different constraints to a fixed engine). The first application of steam-power to maritime transport is often attributed to Denis Papin, in 1685, but he only devised the principle (see p. 000). The first to submit a specific patent was the Englishman Jonathan Hulls in 1736; it consisted of a tugboat powered by paddle-wheels. The Frenchman J.-C. Perrier was the first to experiment with a steamship, which he did on the Seine in 1775; this was the ancestor of all steamships, although the results do not seem to have been very convincing.

The first conclusive results were obtained by the *Pyroscaphe* made by the marquis

The hazards of steam-power meant that even with powerful engines, metal hulls and propellers, the boats kept 'auxiliary' masts until towards the end of the 19th century; indeed, ever since the transatlantic crossing by the *Savannah* in 1819, which could use her engines for only 85 hours and which arrived in sight of Ireland having burned all her coal, sails were considered an indispensable security. The first large-tonnage vessel to be built without sails was the Peninsular & Oriental's *Himalaya*, dating from 1853; its speed in the experiments was 13.9 knots.

Steam engines were to increase their output right throughout the 19th century, owing firstly to the **compound expansion engine** invented by the American James P. Allaire in 1824 (despite his failures at first). In this type of motor, after the steam had been used in the first cylinder, it was used again in a second cylinder at a lower pressure, thus significantly improving the output. In fact, on the *Brandon* which was the first ship to be fitted with this kind of motor, the relation of 2 to 2.5 kg of coal per horsepower fell to 1.75 per horsepower. The supremacy of the steam engine in maritime propulsion was assured by three developments: the adoption of **tubular boilers**, where the water passed through tubes in the fire instead of the inverse; the **increase in pressure**, which enabled engines four times larger to be made; and especially by the advent of **steam turbines** invented by the Englishman Charles Parsons, which enabled speeds of 35 knots to be reached.

Claude de Jouffroy d'Abbans in 1783; it too consisted of a ship with wheels and a wooden hull. For the first time the possibility of adapting steam to large-tonnage ships was envisaged, and the Englishman William Symington set out with this aim for his tugboat *Charlotte Dundas*, which was powered by a two-cylinder steam engine activating the wheels. In a demonstration this vessel effectively pulled two 70 t barges over a distance of 32 km at a speed of more than 3 knots. These pioneers were separated by three-quarters of a century from the final adoption of steam by the navy.

A special place in the story of the marine steam engine should be reserved for the American James Rumsey who in 1787 invented, made and tested the **first maritime jet engine**: this consisted of a craft which, using a pump, sucked water in at the prow and forced it backwards to the stern. After being tested on the Potomac it hardly had any future at all, for it was very bulky.

In 1803 Rumsey's compatriot Robert Fulton, a brilliant adapter rather than a real inventor, who had a great interest in **submarines** (see p. 227), built a steam barge which he launched on the Seine; the boat had been built with the help of the United States ambassador in Paris, Robert R. Livingston, and it was propelled by an engine specially designed by Fulton. The vessel did not arouse much interest and so Fulton took his projects up again with Symington for the purposes of the English navy. Since the victory of Trafalgar in 1805, however, the British admiralty were no longer interested in maritime innovations, and after being away for twenty years, Fulton went back to the United States where he built a new boat called *The Steamboat* which was 40.5 m long and could carry 100 t. In 1807 this boat achieved the historic feat of going from Albany to New York, therefore covering a distance of 390 km, in 62 h at nearly 4 knots, which did arouse considerable enthusiasm. It could be said to have been *The Steamboat* which won the cause for navigation by steam, at least in certain conditions; in fact, however, being a boat with wheels like the *Phoenix* which Fulton built immediately afterwards, it fared badly in bad weather.

It was not until several years later that the propeller and the metal hull together were to provide steam with the real basis of a revolution in maritime engineering.

Travel agency

Cook, 1841

The travel agency undoubtedly constitutes a major sociological and commercial invention, for it has led to the creation of the **tourist industry**. It was started up by the English pastor Thomas Cook in 1841. Cook persuaded a railway company, the Midland Counties Railway Company, to run a special train service between Leicester and Loughborough on 5 June of that year for the transportation of the participants in a conference on temperance. These were the first special trains in history. In 1844 the same company proposed to Cook that they would keep the service going if he could guarantee them a sufficient number of passengers. In 1855 he organized special trips between Leicester and Paris where a Universal Exhibition was being held, for which he acted as guide. In 1856 he extended this formula to a grand tour of Europe. It was only in 1861 that Cook really founded the very first travel agency which sold train and boat tickets and took care of hotel reservations.

Tyre

Thomson, 1845; Dunlop, 1888; Michelin, 1895

The first person to submit a patent for a **rubber tube** filled with air fixed to the rim of a wheel, which offered a certain amount of traction at the same time as cushioning impacts, was the Englishman Robert W. Thomson in 1845. The first Thomson tyres, which consisted of inflexible casings around an **inner tube**, were designed for vehicles pulled by animals; they were very successful, but were superseded nevertheless by **solid tyres** by which they were ousted after a few years. The newly successful bicycle inspired the Scottish veterinary John B. Dunlop to patent some bicycle tyres in 1888. The Michelin brothers were the first to fit a motor vehicle with **tyres with inner tubes** in 1895.

> The first tyres were made by Thomson and were called '**air wheels**'. They broke an endurance record of some 2000 km when they were fitted to the wheels of a brougham.

Underground railway

Pearson, 1843; Bienvenüe and Chagnaud, 1898

The growing congestion of 19th-century urban traffic prompted the councillors and engineers in the large towns of the world to look for ways of decongesting it. In New York **elevated railways** were built, but these just aggravated the pollution and the din. In 1843 the Englishman Charles Pearson was the first to suggest the building of underground tunnels in London through which railway lines could be laid. The project was not approved until 1853 and construction did not begin until 1860. Despite the **sulphurous gases** emitted by the system of re-injecting the exhaust into the steam locomotive boilers, the **London Underground** was immediately successful, for it transported some 10 million passengers in its first year of service in 1863. While the London network expanded and became electrified (1890), Glasgow began to build its own in 1886, followed by Boston, Budapest and Paris (1898). The construction of the Parisian metro was speeded up as a result of a technique invented by Fulgence Bienvenüe and Louis Chagnaud, which consisted of digging the tunnels as open-air sections, strengthening the vaults and then roofing the trenches. This method of cutting sections of trenches caused considerably fewer problems for the traffic and the navvies.

> When the New York metro was built in 1912, the navvies were surprised to happen upon a section which had already been made, with a magnificent waiting room. This had been dug in 1870 at the personal cost of Alfred Ely Beach, co-editor of the *Scientific American*, and then forgotten. Beach had hoped that the New York town council would approve his **pneumatic train** project, but some scandal hindered the completion of this project although it was well underway.

Wheel

Anon, Ur, 3500 BC

The oldest known representation of a wheel to have survived is that on a tablet of clay from Ur in Mesopotamia, which dates back to 3500 BC; it consists of a **solid wheel**, since the first **spoked wheel** did not appear until around 2000 BC in Egypt, and then in Syria. Strangely enough, the wheel seems to have been unknown in most of the pre-Colombian civilizations.

It might seem that the use of wheeled vehicles expanded continuously throughout history, but this is not the case at all. On the contrary, these vehicles tended to disappear from numerous regions of the world, because animal transport and, particularly in the Near and Far East, **camel transport**, provided very effective competition. In the 3rd century for example, at the start of the Sassanid dynasty in Persia, wheeled vehicles had almost completely disappeared, and from the year 30 BC in Egypt, the camel became increasingly competitive. Some modern experiments have shown that this regression was not illogical: in many regions the paths were often very rough, and so the camel provided transport that was quicker and more profitable. At the time of the Muslim conquest, wheeled vehicles had almost completely disappeared from the Near East and they only reappeared, in Syria and Egypt for example, in the 19th century when the roads were rebuilt.

Warfare

If military technicians had put their inventive genius to good use in other domains, the destiny of humanity would undoubtedly have been improved. Apart from mechanical locomotion and the atomic bomb, very little exists that was not invented during the first few centuries, if not before. Incendiary bombs, gas warfare, flame-throwers, and bacteriological and even biological warfare were very successful in China early in history. In the 10th century the Scandinavians had armour-plated boats with which they rammed the unfortunate, unarmed vessels of their enemies. The gun had been made by the 15th century, and its technological progress seems endless.

This war fever comprises some colourful chapters. One of these is the invention of the submarine, which in its youth proved to be as dangerous for the sailors as for the enemies, and which was also very tiring since the propeller was turned using a manual crankshaft!

Another curious chapter is the invention of armour-plating for warships. At the time these ships were made of thick wood in which cannonballs could only make small holes, but this was quite enough for the admiralty and the sailors, in fact, for everyone. Crash! The Russians tested the damage that could be inflicted with their new high-explosive shells, and Dupuy de Lôme, pioneer of battleships, was at last urgently summoned in 1858 to build the floating fortresses that he had been predicting for ten years.

Fortunately, however, no one ever again found themselves in the position of the American Indians being attacked by the Spanish conquistadors, when the Indians were thrown into total confusion by the firearms and horses of the invaders. After just a few years technological progress has now been shared among everyone. Death too.

Armour-plating (of warships)

Anon, Scandinavia, 11th century; Congreve, 1805; Stevens, 1812; Dupuy de Lôme, 1848

The idea of armour-plating vessels may date back to Archimedes who in 250 BC had a ship built for king Hieron of Syracuse, the sides of which were protected by a mesh of ropes supported by iron chains. This ship was the *Syracusan*, the first ever to benefit from a reinforcement of its hull against enemy attack. The Romans developed protected vessels, such as the **quinqueremes** or galleys with five rows of oars, which had a covered deck to protect the rowers against enemy arrows and missiles. The Byzantines took up this idea again but modified it, replacing the wooden deck, which was heavy and did not allow much ventilation, with sheets of thick leather. The first armour-plating, in the modern sense of the word, seems however to have been invented by the Scandinavians, who armed their vessels with prows and iron plates in the 11th century — a procedure which was adopted at least once in Catalonia in the 14th century for the protection of a ship which seems to have been terrifying, at least for the time.

At the beginning of the 19th century the great thickness of the hulls — some 50 cm of oak wood — made the warships reasonably resilient to **cannonballs** fired from a distance, and all the more so to **bullets**. The cannonballs fired at short range usually only succeeded in piercing a few planks, occasionally causing loss of life, but only very exceptionally sinking the ship, unless the ammunition store was hit or the bullets made holes at the water-line.

Two technical inventions changed naval military engineering: the advent of **steam** for propulsion and the invention of the **high-explosive shell**, which was to be tested for the first time in 1837 (see p. 223). One year before his rocket was made (see p. 225), the Englishman William Congreve, one of the most brilliant military technicians of all time, proposed to build **floating batteries** of mortars under armour-plating, which, as he wrote in his letter to *The Times* on 20 January, 1805, would withstand any artillery. It was a premature proposition and was not realized. In 1812 the American John Stevens suggested to Congress the construction of armour-plated ships. The marines were sceptical, however. Senior management and sailors were united in their continuing confidence in protection by thick wooden hulls. Their scepticism was rudely tested however when in 1837 the English fleet, having adopted the formula of the French general Henri-Joseph Paixhans, destroyed the Mexican fortress of San Juan de Ulloa using high-explosive shells. In the following years several English, French and American engineers developed projects of steam-powered **armour-plated frigates**, which did not succeed because the cause of armour-plating had not yet been won. In 1854 the Stevens family undertook to build a 4200 t vessel with a 38 mm steel belt round it, a vessel which was never finished. The obstinacy of the military of the time is comparable to that of their successors in the 1920s who would not believe in the role of aviation in a battle. The Frenchmen Gaston Gervaise and Dupuy de Lôme, the same

Armour-plating gave blatant proof of its superiority during the American Civil War. On 8 March, 1862, the battleship *Virginia*, flying the Southern flag, burst in on the Northern fleet which had organized the blocking of the York and James Rivers at Hampton Roads; it rammed the *Cumberland* and threw the *Congress* on to its side, despite a volley of ordinary bullets. The very next day the Northerners sent an entirely new ship against the *Virginia*; this was the *Monitor*, a battleship without masts built by the Swede Ericsson and equipped, for the first time, with two guns placed in a 140 t **armoured gun turret**. None of these battleships sank and despite having the advantage of its tower, the *Monitor* did not win; its seaworthiness was only mediocre but it gave proof of the importance of armament in the new armour-plated fleets.

Dupuy de Lôme who was later to be one of the makers of the submarine the *Gymnote*, were therefore champing at the bit.

Then, like a thunderbolt, on 30 November, 1853 the 'Sinop affair' broke out: in less than two hours the Russian fleet, which was two frigates and six ships strong and under the command of Admiral Nakhimov, destroyed at anchor the Turkish squadron of Admiral Osman Pacha, which was seven frigates, three corvettes and three steamships strong, using high-explosive shells. The cause for armour-plating was won and the clan of sceptics was annihilated several months later on 17 October, 1854, when the Franco-British squadron was severely damaged by the Russian high-explosive shells at the siege of Sebastopol. This disaster signalled the end for the wooden ship. With great urgency the English and the French undertook the construction of battleships beginning, in France, with five 1600 t '**floating batteries**', plated with iron 10 cm thick. Dupuy de Lôme, who had become director of naval shipbuilding, then began the building of a ship which had been designed ten years previously, the 5600 t, 78 m-long, armour-plated frigate *Glory*, which had a 900 cc engine. Its plating alone weighed 840 t, and to compensate for this excess weight Dupuy de Lôme reduced the artillery to thirty-six 160 mm cannons which fired high-explosive shells. This was the birth of the battleship. Five others followed: the *Crown*, the first to have a hull made totally of iron, the *Normandy*, the *Invincible*, the *Magenta* and the *Solferino*.

Bacteriological and biological warfare

Leonardo da Vinci, around 1500

Although **microbes** were not recognized as disease carriers until the end of the 19th century, their military usage has been hinted at since the beginning of the 16th century. It was actually around 1500 that Leonardo da Vinci planned to bomb the enemy with missiles containing a liquid extracted from the saliva of a mad pig or dog. It was also Leonardo who suggested incorporating **biological poisons** such as **venoms** from toads and tarantulas into bombs, but the plan never came to fruition.

Bacteriological warfare was envisaged at least once during World War II. Around 1942 plans for bombs containing the fearsome **anthrax bacillus** were developed. The Scottish island of Gruinard, where the experiments took place, was for a long time forbidden to visitors because of the ground being toxic, but it has recently been declared safe.

Ballistics

Tartaglia, 1537; Blondel, c.1675; Varignon, 1704 and 1707; Bélidor, 1739; Robins, 1742; Bernoulli, around 1745; Euler, around 1753

Ballistics, or the **study of missile trajectories**, developed from **mechanics**, and for a long time it suffered from erroneous theories on movement. Thus in the 16th century it was thought possible to distinguish between slow movements and fast movements, a distinction which cannot be upheld in ballistics. When the Italian Niccolo Tartaglia published the first concepts of ballistic theory in his *New Science* in 1537, he added to existing incorrect ideas one which was just as false, that of a maximal range at an angle of 45°. This was not based on any mathematical information and for him it constituted an average value between **vertical shooting** and **horizontal shooting**. Paradoxically, it can be said that Tartaglia invented the concept of ballistics without any understanding of it.

The first to solve the problem did so by eliminating the idea of slow and fast movement, by demonstrating firstly that the **oscillation time of a pendulum** is independent of its range. Newton's universal law of attraction was not yet known; this would have enabled the understanding of the fact that the range of a missile depends on its initial speed and on the quotient of its mass by its frontal surface. Torricelli, Mersenne and Blondel, who were studying ballistics, would have at least established that the trajectory of a missile is **parabolic**.

The Frenchman François Blondel, who was studying the subject around 1675, oddly enough took little notice of **air resistance** but stressed the importance of **initial speed**, since he was interested in the quality of the gunpowder, as well as the **weight** of the projectile.

> Benjamin Thompson, the English count of the Germanic Holy Roman Empire who invented filter coffee making, is sometimes considered to be Einstein's forerunner, for he established the equivalence of heat and energy.

The two memoirs published in 1704 and 1707 by his compatriot, the mathematician Pierre Varignon, outlined the notions concerning the parabola, but had little practical significance. Nevertheless the **bullets** were still fired using gunpowder. Varignon eventually turned to the study of air resistance, which the Englishman Isaac Newton had also done when he took a brief interest in ballistics in 1679–1680. Newton had postulated that the air resistance is proportional to the square of the speed. Varignon took up this theory, as well as the two-term function for the calculation of speed, which follows on from it and implies that a projectile moves with one speed up to a certain point of its parabola, then with another beyond this point.

The problem was that until then no one had really been able to determine either the trajectory or the speed of a missile fired from any type of weapon. In 1739 Bernard Forest de Bélidor had tackled the problem of the **combustion of the gunpowder** and the energy which a given load of powder imparted to a projectile of a certain size, and that year he published his results. Alas! these were not much use at all for they had been deduced from experiments carried out using **mortars** which were different from the ones used by the army. Moreover, there were many varieties of combustion chamber, and so these results could offer no more than indications. There was no question of standardizing the construction of weapons (see p. 137).

Around 1745 the calculation of the parabola was refined by the Swiss Jean Bernoulli, and around 1753 his compatriot Leonhard Euler, a great specialist in curves, introduced a term proportional to the fourth power of the speed into the air resistance calculations. In the meantime, in 1742 ballistics benefited from an invention by the English physicist Benjamin Robins: the **ballistic pendulum**. This was actually simply a big metal screen suspended from an axle on which any impact printed a number of oscillations in proportion to its power, but it was the first known instrument to enable the measurement of the energy value imparted by different gunpowders. Using this pendulum, Robins demonstrated that Newton's law is only viable for projectiles which have a speed greater than 300 m/s.

Robins had a practical mind as well as an acute sense of observation and he made great progress in ballistics. In the first place he used mathematics as a tool for the representation of information, thus liberating this science from abstract speculation. Then he was the first to notice that the projectiles deviated from their path, and to understand the reason for this phenomenon: the friction of the missiles in the **bore** of the gun, which caused them to have a helicoidal movement. These observations were to alter ballistics considerably when applied to the use of the pendulum, for they had three reasonably rapid consequences. The first was the adoption of **grooved bores**; the second, the adoption of **elongated missiles**; the third, the **loading of missiles from behind**, with the grooves of the bore stabilizing the trajectory of the missile. A fourth consequence later arose: the need to calculate the length of the gun in proportion to the optimal trajectories and type of missiles.

Euler's prestige was a good deal greater than that of poor Robins, and unfortunately he rejected Robins's ideas, because they contradicted his intellectual beliefs. Euler's genius is not to be questioned here, for he proved it in numerous other sectors of mathematics, but it must be noted that his stubborn opposition was paid for by a century's delay in the progress of ballistics! Perhaps Euler was unwilling to accept that Robins had relegated mathematics to a simply representative role, and, what is more likely, perhaps he did not know that the problem of **deviation** had been recognized in the 17th century and that it had been resolved, at least in rifles, by the making of guns with grooved bores.

In the domain of ballistics homage must be paid to a certain Jean-Charles de Borda who, in his memoir of 1769, tempered the excessive role of mathematics in the calculation of parabolas, quite independently of the observed facts. He therefore introduced the principle of approximation, which was certainly empirical, but less gratuitous than pure mathematical calculations. Homage must also be paid to Jean-Baptiste Vaquette de Gribeauval, also known for his role in standardization in weapon manufacture, who — not without bitter dispute — introduced the use of the **backsight adjuster**, a device which was fixed on the top of the gun to facilitate the **aim**, and which, according to Gribeauval, also meant that one aimed above the **line of sight**. Even with the backsight adjuster, precision in shooting left so much to be desired that in 1816 people were advised to practise estimating distances by sight!

Not only, therefore, was firing inaccurate, but the marksmen were steeped in empiricism. Enormous problems remained to be resolved before ballistics could become a science worthy of its name. The most difficult of these were the **laws of the power of the gases** released by the combustion of the gunpowder and the **elas-**

At the time of Napoleon's campaigns, **mortar shooting** was still carried out at 45°, in defiance of the laws established by Bernoulli which indicated that the maximal range was attained at smaller angles. No one knew where the missiles fell; adjustments were made after observation of the explosion through binoculars, then the trajectory altered by changing the charge of gunpowder. Although the artillery-men had acknowledged the principle of the air resistance being proportional to the square of the speed, tradition still had the power of law and no one had thought of making gun carriages which would allow firing angles greater than 6° to be obtained with field guns. With an angle of 4°, a range of 1500 m could be attained, with 8°, 1650 m and with 12°, 1800 m. The fight therefore took place at quite close quarters. Had Robins's notions been studied in depth, particularly the need for guns with grooved bores, then France, for example, could have conquered Europe. (It must be noted that this information constitutes only a summary of the inventions which have contributed to the state of modern ballistics, of which there were a great many.)

ticity of the carriages, which were the result of **interior ballistics**. Numerous and often famous people became interested in this new domain, that of the trajectory of a projectile, which was henceforth called **exterior ballistics**. These included Poisson, Legendre, Gay-Lussac, Daniel Bernoulli and Bigot de Morogues. It was only in 1797 that Benjamin Thompson, Count Rumford, finally gave the formula which linked the **explosive force** of the gunpowder to the **density of the charge**.

In 1844 the French navy engaged in the construction of a **grooved bore gun**, the first of its kind, which took nearly ten years. This gun, a 16 cm howitzer cannon, was not ready until 1855, just in time for the siege of Sebastopol. In the meantime, England and Italy had recognized the brilliance of Robins's ideas and had also begun to apply them.

Cannon

Archimedes, 2nd century BC

The idea of using a tube to propel a cannonball seems to go back to Archimedes; it would have been described by him in one of his works which has not survived to modern times. In any case it was to Archimedes that Leonardo da Vinci attributed the invention of a steam-driven engine of war. According to the description it was a steam cannon, consisting of a tube, perhaps of copper, the end third of which was heated before the addition of water, which then turned into steam. It is not known by which mechanism the steam would have pushed the cannonball, which weighed, according to Leonardo, one and a half talents, or 36 kg. It is doubtful that this invention was ever used, especially on a large scale.

The invention of the gunpowder cannon itself is anonymous, and seems to date back to the second decade of the 14th century. The gunpowder cannon is cited as appearing at the siege of Metz (1324) and Cividale, in Italy (1327), but the only definite date is that of 1346, when the English king Edward III used cannons at Crécy. The weapon does not seem to have affected the result of this battle, nor that of subsequent ones until the beginning of the 18th century. It was the perfecting of bronze metallurgy which eventually caused the cannon to come into widespread use, pushing the range from 300 m/s to 1000 m/s. The invention of the grooved bore by La Hutte in 1858 increased accuracy and doubled the range of fire.

Archimedes's ***steam cannon*** *sketched by Leonardo da Vinci*

Gas warfare

Anon, China, 4th century BC

It is difficult to know whether to praise or deplore Chinese technological genius on the subject of gas warfare, of which they were undeniably the inventors. Indeed, during the 4th century BC they used **bombs** and **grenades** based on dry **lacquer**, a powerful lung irritant, **arsenic** and **lead oxides**, incorporated into a combustible base such as **resin** or **wax**. The bombs and grenades were set on fire before being thrown towards the enemies, spreading debris among them which gave off toxic vapours. At this time they also employed an original technique consisting of digging tunnels just below the enemy camps and diffusing poisonous fumes into them.

The technology continued to improve over the centuries and the invention of **gunpowder** provided the inspiration for particularly toxic explosive bombs, based on dried and finely ground **human excrement**, **aconitine** — an extract of aconite and a very strong alkaline which, even in weak doses, causes paralysis of the respiratory muscles — **croton oil**, equally toxic, **arsenious oxide** and **arsenic sulphur**, and crushed **cantharid fly**, as well as plant seeds releasing a strong smoke which prevented the enemy from opening their eyes. The toxic missiles, called '**magic smoke apparitions**' were propelled from a piece of artillery and were particularly effective when they fell inside the fortresses. The concoctions were very varied and made use of all the known animal and plant poisons in China, including dried **mustard seeds**, the smoke from which contains the highly toxic esters of **isothiocyanic acid**. In the 17th century the Chinese, who were still practising gas warfare, changed their preference from arsenic to **mercury derivatives**, which have lethal effects on the nervous system.

The Chinese are also responsible for the use of substances such as **napalm**. In the 15th century they filled their bombs with a substance which ignited on explosion and stuck to the skin.

Finally, they used **tear gases**, which were spread among the enemy lines by flaming torches attached to horses' tails.

Chinese howitzer *of the Middle Ages, which 'erupted with fire to the sound of celestial thunder'. It dispatched projectiles emitting toxic gases to the enemy*

Gas warfare was first used in modern times during the 1914–1918 war, and since then has made only a few brief reappearances, in the Middle East during the 1980s.

Gun

Ctesibius, 245 BC; anon, Arabia, 1304; Marin Bourgeois, c.1498 etc

The first mechanical weapon which evokes the gun seems to have been the **compressed air blowpipe** which was made in 245 BC by the Greek Alexandrian engineer Ctesibius. It consisted of a bronze tube in the breech of which there was a piston; this was pulled back against a soldered partition until the air reached its maximum compression. This may be called the **breech chamber**. The piston was then stopped by a **stop-catch**. When this was pulled, the compressed air was released violently, forcing out various projectiles, usually arrows. The range of such a weapon could have been around 50 m. The invention does not seem to have been followed up. The following stage in the invention of the gun was the **gunpowder blowpipe**, which was used in Arabia from 1304. It would have consisted of a segment of bamboo lined with copper, which was of varying reliability. This invention had no immediate follow-up either.

The compressed air blowpipe nevertheless reappeared in France around 1498. It was made by Marin Bourgeois of Lisieux, and was drawn and reproduced in 1608 in David Rivault's *Eleménts d'artillerie* (Artillery Equipment). The weapon comprised a copper breech and an iron barrel; it was 1.35 m long and worked on exactly the same principle as Ctesibius's blowpipe. It fired **metal or wooden balls** with iron points and would have had a range of 350 m. There may have been another version of this, made in Nuremberg in 1570 by Guther, but it was no more successful.

Twenty-two years before Bourgeois's compressed air blowpipe appeared, an improved model of the Arabs' powder blowpipe was tested at the battle of Moret (1476): this was the **arquebus**, a very heavy weapon which needed two men to carry it, and which could only fire once every five minutes, provided that it was not raining. It is said to have been invented in Spain around 1450, but its inventor was actually an adapter who changed the bamboo tube to one of metal and placed it on a butt. In 1503 the arquebus proved its efficacity once again by flattening the French and the Swiss mercenaries at Cerignola. The firearm triumphed over the pneumatic weapon, therefore, although it had about a third of the range. Around 1570 the **musket** was to replace the arquebus, despite it weighing even more; this weapon, though actually not fundamentally different from the arquebus, had a range which was a third greater: up to 300 m. At the beginning of the 17th century the Germans improved the **ignition system** by replacing the manual ignition using a fuse with **a flint ignition**; the fuse was coiled in case it had to burn for a long time. This weapon, the flintlock musket, was the first to be referred to by the word 'gun'. The German improvement, with which the wheel lock musket competed, was in fact derived from the spinning wheel arquebus, where a steel **spinning wheel** produced a spark to light the fuse by knocking against a flint. In the 17th century the musket gun, chosen, for example, by Vauban, differed from its pre-

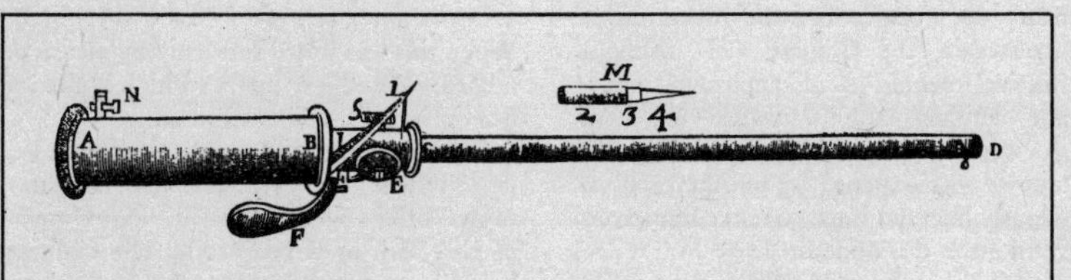

Marin Bourgeois's 16th-century ***air gun****, as featured in* Eleménts d'artillerie *by Rivault, 1608*

decessors in only one significant way: its bore was reduced to 18 mm.

The modern gun is the product of a series of secondary improvements, such as **percussion ignition** (dating from 1807) which, due to a **fulminate cap**, avoided the misfires which occurred when it rained; this was the work of the Scottish minister Alexander Forsyth. At the same time the first experiments with loading via the breech took place, to which it is difficult to attribute a single inventor; it is known only that the **stopper breech** was invented between 1835 and 1841 by the German Johann Nicklaus von Dreyse.

Though it was studied in the United States in 1854, the repeater rifle only appeared in 1860, under the name of the **Henry rifle**; in 1866 an improved model was named the Winchester, after the industrialist Henry Winchester who had bought Smith & Wesson's firm Volcanic, where the Henry rifle had been developed.

Gunpowder

Anon, China and India, 9th century

The oldest known mentions of gunpowder are those found in Chinese and Indian manuscripts dating from the 9th to the 11th century: a mixture of **saltpetre**, **sulphur** and **charcoal** is described. This mixture was intended for use in fireworks. The first known military use dates back to the 13th century (see *Gun*).

Gunpowder, as used by the military in Europe, remained until the 19th century the only known **explosive** used in the civil domain. It was not, however, until the 17th century that it first came into use, for the building of tunnels in France. It was superseded by **nitroglycerine** (see p. 29) and **nitrocellulose**.

Land and sea mines

Anon, China, 1277; anon, China, 14th century; Colt, 1843

The first known mention of a land or a sea mine, that is, an explosive device not propelled by a gun or any other means, has been found in Chinese historical records. This is not surprising for since the 2nd century BC, even before the invention of **gunpowder**, the Chinese were using a primitive technique of explosives (they made bamboos explode by throwing them into furnaces) and the mastery of gunpowder, sharpened by military genius, obviously inspired them to make numerous inventions in this domain.

In 1277, having made much progress in the technique of bomb manufacture, the Chinese, apparently for the first time, thought of burying very large bombs in the ground which they could detonate from a distance using a process which was not fully understood until the 17th century: it consisted of placing a **detonator** made of fragments of flint and iron — almost certainly a device on a wheel — inside the bomb. When this was jolted suddenly by means of a cord, it produced sparks which ignited a fuse, possibly tow soaked in oil.

Some underground bombs were enclosed in spherical iron cases just like the later mines; others were lodged in bodies made of porcelain or baked earth. The Chinese had also invented **deep land mines**; these were lodged inside 2.70 m-long hollowed-out bamboos which were then driven well into the ground. The explosive part was

Chinese ***mine-layer barge****. A late model, since it dates from the 16th century, it is very sophisticated: its front section is put to drift as soon as the oarsmen have approached the enemy target and have ignited the fuses of the mines. The back section then returns nimbly to harbour*

placed at the bottom end of these tubes.

Mines such as these were put down in fields by the hundred in order to protect narrow passes, outskirts of towns and entrenchments, just as in contemporary warfare. This defensive system, described by a British manual in 1412, was not adopted by Europeans until the 16th century; it was the German Samuel Zimmerman who used it for the first time, in 1573, keeping the same mechanism for detonation at a distance as the Chinese had used.

The **sea mines** came a short while later, appearing in the 14th century, still in China. Smaller in size, they were inserted inside bovine bladders and balanced in such a way that the fuse emerged from the surface of the water; the combustion time of the fuse determined the time limit between the launch of the mine and its explosion. These mines were cleverly fixed to planks of wood and conveniently ballasted so that they floated a little below the water-level. They were launched in the direction of the current towards the enemy fleets; the combustion time of the fuse had to correspond to the time it would take to drift towards the enemy ships. This method was suggested by the Englishman Ralph Rabbards during the reign of Elizabeth I and it must have been effective because the Chinese used it themselves in their fights against the English gunboats on the rivers in 1856.

In 1843 the American Samuel Colt made a type of sea mine, the explosion of which he controlled from a distance using a **submerged electric cable**. In this way he managed to make a mine explode and sink a ship from a distance of some 8 km. The principle was adopted by the Prussians in 1848 during a war against Denmark, but it was abandoned in 1900.

Machine gun

Anon, 14th century; Vinci, c.1480; Puckle, 1718

The idea of a machine gun, that is, a light weapon capable of firing several shots in succession, dates back to the 14th century. At this time weapons appeared which consisted of guns arranged one next to the other on a chassis and which were ignited successively, or perhaps at the same time if there were enough battery attendants. They originated almost certainly in China, for it was at the same time that the Chinese first made use of **flame-throwing organs**, which consisted of ten guns activated by ten servants. Around 1480 Leonardo da Vinci sketched several of this type of organ, one of which comprises 33 guns arranged in rows of eleven and attests to definite theoretical progress; they were placed in a fan shape which allowed simultaneous ignition. It seems, however, that their success was limited because they were too cumbersome.

In 1718 the Englishman James Puckle made a remarkable technological leap to the type of weapons which were called **ribauds** or **ribaudequins** and which had been adapted, furthermore, as hand weapons (see p. 224); the most sophisticated were equipped with devices such as a pivoting firing pin, which enabled the successive ejection of all the bullets from the guns — but that presupposed that the trigger was pulled each time. Moreover, as with the first revolvers, ignition using the only available means — the flint — could sometimes cause misfires or all the charges to go off at the same time, making the organ or ribaud explode. Puckle thought of a **pivoting loader** of nine bullets which would equip a light gun mounted on a **tripod**. The pivoting loader was borrowed from the revolver, and the chassis was almost identical to that of the previous machine guns. This weapon was tested, judged operational ... and was mysteriously forgotten.

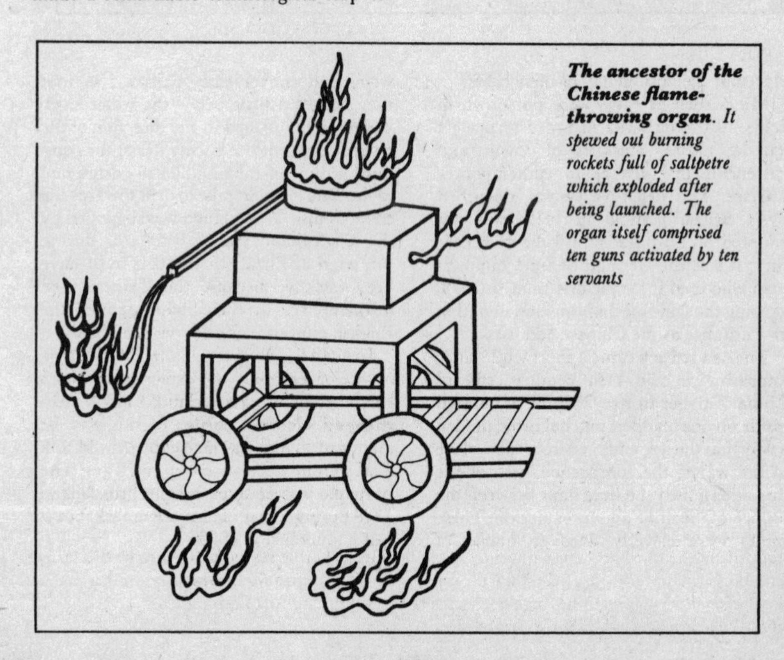

***The ancestor of the Chinese flame-throwing organ**. It spewed out burning rockets full of saltpetre which exploded after being launched. The organ itself comprised ten guns activated by ten servants*

Marine high-explosive shell

Anon, Venice, 1376, or Netherlands, c.1370; Deschiens, 1690; Bentham, 1788; Paixhans, 1822; Vieille, 1884–1886

The invention of the high-explosive marine shell is one of those with a sporadic history. The **explosive military projectile** had existed in the form of the **grenade** since at least 777 (see p. 218). The **exploding cannonball** had been invented to replace the old **solid cannonball**; according to some authors it dates from Venice in 1376, and according to others it was invented around 1370 in the Netherlands. The device was certainly rudimentary and dangerous for the operator: it was a hollow sphere made of iron, filled with **black gunpowder** and pierced with a hole through which it was filled; it had a fuse which was lighted separately, before the cannon itself was ignited. It was not used with regular frequency: the exploding cannonball (which was not yet a **cylindro-conical shell**) appeared and disappeared from the battlefield, depending on the strategies and the tacticians who did or did not have confidence in it.

There was still no question of using it in the sea. The reason is obvious when one considers that until 1881 the sailors practised **loading by the mouth** of the cannon. If the high-explosive shell happened to get jammed in the barrel, then the whole ship would be in danger. Moreover, the explosive cannonball could lose some of its powder or its lighted fuse as it rolled down the barrel. In 1690, a Frenchman, Captain J.M. Deschiens, contravened all previous practices: he loaded the high-explosive shells by the mouth of the long bronze cannons and then fired at once (horizontal position). The shells were his invention, as was the technique. In 1690 he single-handedly put four British ships to flight, then sank a Dutch ship, and damaged another. It can therefore be said that Deschiens is the inventor of the nautical high-explosive shell and its method of operation. He was too much ahead of his time, however, and could not single-handedly overcome the sailors' prejudice against the high-explosive shell. Nearly a century later the Englishman Sir Samuel Bentham, who occupied a top position in the Russian navy, either reinvented or rediscovered Deschiens's technique. He put brass cannons on launches and fired incendiary cannonballs at the Turkish fleet, which were dispatched to the bottom of the sea.

In 1803 the Englishman Henry Shrapnel had the idea of filling a spherical cannonball with musket balls embedded in black gunpowder which would be ignited using a wick fuse. When this set the powder on fire, the cannonball would burst and release the balls. The inventor subsequently made his projectile, which was henceforth known under the name of **shrapnel**, into the cylindrical-nose cone shape of a shell. In 1807 the Scot, Alexander John Forsyth, invented the **percussion cap**, bringing to realization the first real high-explosive shell: a delay fuse at the base of the projectile caused an explosion at a determined point in the trajectory.

During the 1850s, some time before the completion of the actual **percussion shell** — where the explosive cap is placed at the top of the nose cone — the French general Henri-Joseph Paixhans grasped the advantage of using high-explosive shells in naval warfare and in 1822 he equipped **steamships** with weapons which were designed for this type of missile.

All the world's sailors then had food for thought during the **battle of Trafalgar**: not one ship was sunk due to artillery. It was another 'battle of Trafalgar', the **battle of Sinop**, which eventually proved the importance of nautical high-explosive shells to the admiralties and alerted them to the need for **armour-plating for battleships**. Moreover, in 1853 Admiral Nakhimov's fleet had sent most of the Turkish fleet to the bottom of the sea using high-explosive shells. In 1886 the Frenchman Pierre Vieille invented **strong detonating power explosives** which began the age of modern high-explosive shells.

Metal cartridges

Howard, 1799, or Forsyth, 1807; Shaw, 1816

Since the beginnings of artillery, **black gunpowder** was used for expelling the missiles from the barrel of cannons; the powder was ignited by sparks obtained by rubbing a piece of steel on a **flint**. This system remained in force until the end of the 18th century or the beginning of the 19th century. It is not certain whether it was the Englishman Edward Howard or Alexander John Forsyth who discovered the explosive properties of **mercury fulminate**. In 1799 Howard seems to have discovered that the black gunpowder to which mercury fulminate had been added was more explosive, and Forsyth seems in 1807 to have thought of using this substance to make **explosive caps**.

In any case, the first to apply these caps to firearms was the American Joshua E. Shaw. In 1816 he made a chamber, first of tin and then of copper, to protect the black gunpowder from the moisture in the air. Then he placed mercury fulminate in the swell at its centre, at the position corresponding to the point of impact of the **firing pin**. Under impact the mercury fulminate exploded and in turn generated the explosion of the propelling agent, the black gunpowder. For the first time it became possible to dispense with an external flame. During the 1860s this type of ammunition led to the production of cartridges made out of paper crimped inside steel cylinders — cartridges — at the base of which, in a hollow, was the explosive cap of mercury fulminate. Shortly afterwards the American manufacturers began selling bullets for the use of infantrymen which were made entirely of brass, comprising the missile, the propelling agent and the explosive cap. These metal cartridges significantly changed the evolution of machine guns (see p. 222).

Revolver

Anon, England, c.1630; Lenormand, 1815; Collier, 1818; Colt, 1835–1836; Lefaucheux and Galand, 1860–1868

The word 'revolver' is derived from the verb *to revolve* and was used in the 17th century to refer to guns with several barrels turning on an axle. These were flint guns which must have been of only average efficacity, considering the little progress made in metallurgy at the time and the absence of **metal-coated bullets** which two centuries later would enable the extended use of this type of weapon. In 1815 the Frenchman Hippolyte Lenormand took up the principle of these old weapons, then called '**pepper-pots**' — which date back to the 16th century — and replaced the **flint percussion** with **cap percussion**, which ejects the bullet. This model aroused a certain amount of curiosity and was

> The Armament Museum at Ingolstadt in Germany has a weapon which is undoubtedly the oldest known revolver, since it dates from the 1580s; of anonymous manufacture it consists of a cylinder of three tubes soldered around an axle which was controlled manually. It is called a '**spinning wheel revolver**', because it is equipped with a **spinning plate** which arms itself automatically when the hammer is armed, the spinning wheel generating a spray of sparks. This would suggest that, contrary to many common assertions, the English invented the word 'revolver' but not the concept. Weapons such as this are assumed to have been rather unsuccessful since they are extremely rare.

improved in other countries, notably in Germany and in England, but it did not become widespread. Three years later the Englishman H. Collier inverted the principle of the Lenormand revolver and built the first **turning cylinder revolver**. The cylinder comprised five chambers and had to be turned manually. In the years 1835–1836 the American Samuel Colt made another revolver which also had a five-chamber cylinder, but the rotation of the cylinder after each bullet had been fired was automatic. It was heavy, lacked a stop-catch and had a tendency to misfire. It was not manufactured industrially until 1847 and it was only between 1860 and 1868 that the firm Lefaucheux & Galand improved it enough for it to be given the title of military revolver (in fact, its range was only 100 m, and it could only offer satisfactory accuracy at 50 m; its efficacity was often described as psychological).

Rockets

Anon, China, 1232; Congreve, 1806; Hale, c.1850

The idea of the rocket probably began in the mind of the first technician to watch burning arrows flying through the air, a military technique of ancient times. The first historical indication of the existence and use of rockets, however, dates back only to the 1232 siege of K'aifeng, the capital of the Chinese kingdom Honan, by the Mongols. The reference to **flying bombs** which made a terrible noise has caused historians to suppose that the Chinese had discovered both **black gunpowder** and the principle of rockets. The double hypothesis is made plausible by the fact that the Mongols made effective use of rockets at the battle of Legnica in Silesia in 1241, no doubt taught by the Chinese. In 1288 the Arabs attacked Valencia in Spain using rockets, and thenceforth numerous armies adopted and improved the principle of these flying bombs. It is known that the construction of these bombs was rudimentary, and their use undoubtedly hazardous, for they were made of paper which was lacquered or pasted with starch, thus forming a tube which was narrow at one end, the mouth, and was sealed at the front. Relatively speaking, this tube, being loaded with charcoal, saltpetre and sulphur, was the ancestor of **solid fuel rockets**, in which the combustion forced the ejection of a mass of warm air which propelled the whole object.

Around 1650, as seen from the drawings by the Polish artillery engineer Casimirus Siemienowicz, the rocket designs had reached a remarkable degree of perfection, since the principle of **multi-stage rockets** — though still not separating ones — had been discovered, as had that of the **rocket with fins**, the shape of which differs only slightly from the modern rockets which inaugurated astronautics. Remarkable too is the fact that by then even the **accelerating rocket** had been invented, though the name of the inventor is not known. This consists of a main rocket to which back-up rockets are connected. It is

> The history of the invention of rockets, which is rich in avant-garde technological ideas, includes one particularly interesting episode. About 1830 the pyrotechnist Claude Ruggieri, one of a famous family in this field, carried out some launches of rockets with cages of rats and mice attached to them. The experiments had a striking similarity to those done in contemporary astronautics. The cages which contained the animals had **parachutes** which opened at the peak of their trajectory, that is, at a height of several hundred metres, and then came back down to the ground, to the great surprise of the onlookers. Emboldened by his successes, Claude Ruggieri prepared to do the same thing with a little boy using several rockets. He would undoubtedly have carried out his plan if the police had not intervened to forbid this first ever 'astronautical' launch.

not surprising therefore that in 1715 the first military rocket factory in the world was built near St Petersburg by Peter the Great.

Following a century of unpopularity the rocket was revived in India where the Prince of Mysore, Haïdar Ali, made the first metal rocket. This assured a much higher **gas compression rate** than that of previous rockets and consequently a much larger range: almost 1 km. The improvements made in the rockets' manufacture and firing method enabled his son, Tippu Sultan, to cause havoc among the English ranks at the 1792 and 1798 battles of Seringapatam. Tippu Sultan's army comprised a unit of 5000 rocket launchers!

Until then the manufacture of rockets had been somewhat unsophisticated; the first to standardize it was the Englishman William Congreve, around the beginning of the 19th century. A skilled pyrotechnist, Congreve differentiated the **explosive rocket** from the **incendiary rocket** by the composition of the charge, and the length and diameter of the tube. He was the first to think of igniting the **explosive charge** separately from the **propellent charge**, which were placed in different compartments. He established eight different rocket models, the heaviest of which reached 27 kg. These were launched from inclined ramps, for **aerial bombing**, and from horizontal ramps for **field bombing**. Congreve's rockets played a major role in the bombing of Boulogne in 1806, and in the siege of Leipzig during the Battle of the Nations in 1813 and of Danzig in the same year. Copenhagen was razed in 1807, also by English rockets. The English had less success with their rockets during the Anglo-American War of Independence, although they had built a special ship, *Erebus*, from which to launch them; this was the first **rocket launcher ship** in history.

In 1815 Congreve, the true pioneer of modern rockets, improved their ballistic stability by replacing the single ejection mouth, which is the ancestor of the **nozzle**, with five mouths, and he made the first **launching tubes**, out of copper. The range of these rockets reached 3 km, and since they cost the same as and were much more mobile than the mortars, the rockets fared well. Congreve's rockets were so successful that the whole of Europe was incited to acquire them. His models, which were adapted according to the staff in command, resulted in a revolutionary prototype, the **Danish delta-wing rocket**. The modified design of this remains in many types of 20th-century missile.

The ensuing improvement was even more remarkable since it prefigured some 20th-century concepts concerning projectiles, missiles and rockets. It was the work of the British engineer William Hale, and was carried out in 1850. Still with a view to improving stability, he thought of using fins to incline the angles of the gas ejection mouths in order to obtain a **spiralling ejection** which would cause the projectile to spin round and round. This modification was all the more effective since the progress being made in metallurgy at the time enabled rocket bodywork to be made which could withstand pressures of several thousand kg/cm^2.

The constant improvements in artillery competed increasingly with those of the rockets, and the military usage of rockets was in decline until the 1860s. Two successful derivatives appeared during this time, however: the **rocket gun** cable launcher for the shipwrecked and the **rocket harpoon** for whale fishing. The former was invented by the Englishman Henry Trengrouse in 1807, and the latter by the American Thomas W. Roys in 1821. Rockets were no longer used except in a sporadic fashion.

A special place in their history should be reserved for the Russian Nikolaï Ivanovich Kibalchich, for in 1881 he designed the first **rocket-propelled plane** in order to assassinate the tzar Alexander III. It consisted of an unpiloted rocket with fins which was propelled by the combustion of successive charges of explosive candles. This was the direct ancestor of the German **V1** and **V2** rockets which featured in the 1939–1945 war.

The principle was taken up again in 1900 by the Swede Wilhelm Unge who adapted it and christened it the '**flying torpedo**'.

Submarine

Brebbel or Van Drebel, 1620–1624; Bushnell, 1775; Fulton, 1798–1801; Bourgeois, 1863; Dupuy de Lôme, Romazzotti and Zédé, 1888

Legend describes Alexander the Great undertaking an underwater voyage inside a glass barrel, but this appears to be a fabrication; Alexander could nevertheless have gone underwater in a **diving bell**. In 1578 the English mathematician William Bourne followed Leonardo da Vinci by designing, in much more detail, a wooden underwater craft which was covered in greased leather and propelled by oars. The first to build this, and to whom the merit for the invention traditionally goes, was the Dutchman Cornelius Brebbel or Van Drebel. Between 1620 and 1624 he carried out several successful tests on a craft similar to the one designed by Bourne. This small craft, the first submarine worthy of the name, manoeuvred about between 4 and 5 m below the surface of the Thames, and it is said that King James I of England made a brief excursion on board. Brebbel is said to have ensured the survival of his passengers using a gas of his own composition. Considering the vague knowledge of gases at the time, this seems a suspect story. Perhaps Brebbel took waterskins of fresh air on board. His submarine must have been successful, for he built two others and there is no account of any accident.

Strangely enough, this means of travel aroused the imagination to such an extent that in 1727 there were no fewer than 14 kinds of submarine in England alone. However it is already necessary to differentiate the **submarine**, which has its ballasts inside the hull, from the **submersible**, also a submarine, which has external ballasts and which can therefore navigate on the surface in the same way as a patrol boat, for example. It is not known which Brebbel's craft was, but it is known that several of the 17th-century models were proper submarines, the ballasts consisting of skin and leather bags which were filled for the dive and then emptied using a system of twisting bars.

The first military use of a submarine dates back to 1775, to the American War of Independence. David Bushnell built an almost spherical craft for the occasion, the *Turtle*, which could carry only one passenger who acted as navigator, sailor and motor all at once, for this picturesque craft had to be propelled by using a handle to activate a **propeller** which was fixed to a horizontal axis at the front. The only improvement was a pump which enabled the ballasts to be emptied. The *Turtle*'s aim was to blow up a British ship by causing an explosion of a charge of gunpowder, but it was not successful. Bushnell's successor, the American Robert Fulton, did not fail to learn from his experience. In 1797, backing up his proposal with a model, he made the suggestion to the Directoire (for Fulton was an ardent supporter of the French Revolution) that he could rid France of the British fleet which was blockading her coastline. One example of his craft, the *Nautilus*, was built in Paris; it was a 6.5 m-long iron submarine covered in copper. Although it was still propelled manually, in this craft there was the help of a **reduction gearing system**. It could maintain four men under water for four hours. An iron and glass bulb served as an conning-tower. Fulton was so enthusiastic that the *Nautilus* had been built before Forfait, the naval minister, had released the funds. It would appear, however, that the craft did not take

One of the most extraordinary episodes in submarine history concerns the *Sea Devil*, which was built at Kiel in 1856 by the Bavarian Wilhelm Bauer on behalf of the Russians. On September 6th that year, the day of the coronation of Tzar Alexander III in Moscow, Bauer boarded the *Sea Devil* at Cronstadt together with four musicians who played the national anthem under water, while the crew sang ... and they were audible. Despite the 136 dives which this submarine did under human propulsion, the imperial Russian navy rejected it because it found the weapon to be 'deceitful' ...

part in any military action. When the second craft, *Nautilus II*, was launched in Brest harbour heading for the British fleet it was unable to get close to any of the ships because the English, informed either by experience or by spies, had rowing boats on constant patrol around their vessels. *Nautilus II* returned to port without having attached its explosive charge to anything, and France lost interest in the so-called '**fish-boat**'.

Turning to the enemy, in 1805 Fulton tried to persuade England to adopt his submarine. Despite having the support of the Prime Minister Pitt, he came up against the opposition of the first Lord of the Admiralty, Sir John Jervis, who saw in it a serious threat to the British supremacy of the seas. When Fulton succeeded, experimentally, in blowing up the schooner *Dorothy* by attaching to it an explosive charge which was detonated from a distance using an electric cable, the success of his demonstration only served to reinforce Jervis's hostility.

The first military victory by a submarine was carried off by the Southern American captain Horace Hunley and the engineers McClintock and Watson. After building an improved submarine following the Fulton model, which was christened *Hunley* and which was propelled by a **crankshaft** worked by eight men, the three men succeeded in sinking the Northern American *Housatonic* on 17 February, 1864. The *Hunley* sank with it, for its ram had remained stuck in the *Housatonic*'s hull.

Although it had ended tragically, the exploit kindled the imaginations of both inventors and novelists, such as Jules Verne, the creator of the famous, imaginary *Nautilus*. A great many projects and prototypes followed one another. The modern chapter on submarine navigation only really began with the making of a self-propelled submarine, the first example of which seems to have been built by the French engineer Bourgeois in 1863. Measuring 45 m in length, and with a shape which already heralded modern submarines, the *Diver* was propelled by a 80 hp **compressed air engine** (its self-propulsion was reduced by the weak capacity of its tanks of compressed air. Alas! the *Diver* only half dived . . .). Two models of **steam submarine**, realized separately by the English clergyman George W. Garrett and by the Swedish cannon manufacturer Torsten Nordenfelt, were not very successful. In Garrett's model the boilers had to be extinguished before the dive, and so the vessel moved forward only on stored steam; in Nordenfelt's, an improved model, the storage of the steam was problematic.

The first truly automatic submarine was the famous *Gymnote*, made by the French engineers Dupuy de Lôme, Gaston Romazzotti and Gustave Zédé. Although it was launched in 1888 — while a British *Nautilus*, the third of that name and of similar inspiration, had been launched two years previously — the *Gymnote* seems to deserve precedence because its construction began in 1869. It was 17.2 m long, had a diameter of 1.8 m, weighed 2 t and was propelled by an electric motor of 51 hp. This activated a 1.5 m-diameter propeller and enabled speeds which were considerable for the time: 7 knots on the surface and 4.27 knots when submerged. Among the precursors of the *Gymnote*, the *Goubet I* should be mentioned. This was built in 1887 by the Frenchman Claude Goubet, following plans by the Russian engineer Stéphane Drzewiecki d'Odessa, but was rejected by the French naval commission because of its lack of stability. Despite the success of electric propulsion as attested by the *Gymnote*, the principle of the boiler was not immediately abandoned; indeed, it enabled navigation on the surface with the least energy expenditure, and so it was retained by the Frenchman Maxime Laubeuf for the *Narval*, a 34 m-long submarine which was launched in 1899, and which was the successor of the *Gymnote*. The invention of the **diesel engine** a few years later would cause the definitive abandonment of the principle of steam propulsion.

Index

Note: Entries in **bold type** refer to articles headed with the name shown.

DISTRIBUTORS

for the Wordsworth Reference Series

AUSTRALIA

Reed Editions
22 Salmon Street
Port Melbourne
Vic 3207
Australia

Tel: (03) 646 6716
Fax: (03) 646 6925

GERMANY, AUSTRIA
& SWITZERLAND

Swan Buch-Marketing GmbH
Goldscheuerstraße 16
D-7640 Kehl am Rhein
Germany

GREAT BRITAIN & IRELAND

Wordsworth Editions Ltd
Cumberland House
Crib Street
Ware
Hertfordshire SG12 9ET

ITALY

Magis Books SRL
Via Raffaello 31/C
Zona Ind Mancasale
42100 Reggio Emilia

Tel: 0522-920999
Fax: 0522-920666

SINGAPORE, MALAYSIA
& BRUNEI

Paul & Elizabeth Book Services
Pte Ltd
163 Tanglin Road No 03-15/16
Tanglin Mall
Singapore 1024

Tel: (65) 735-7308
Fax: (65) 735-9747

SPAIN

Ribera Libros S.L.
Poligono Martiartu, Calle 1-no 6
48480 Arrigorriaga, Vizcaya

Tel: 34-4-6713607 (Almacen)
34-4-4418787 (Libreria)
Fax: 34-4-6713608 (Almacen)
34-4-4418029 (Libreria)

PORTUGAL

International Publishing Services Ltd
Rua da Cruz da Carreira, 4B
1100 Lisboa

Tel: 01-570051
Fax: 01-3522066

SOUTHERN AFRICA

Struik Book Distributors (Pty) Ltd
Graph Avenue
Montague Gardens
7441
P O Box 193
Maitland
7405
South Africa

Tel: (021) 551-5900
Fax: (021) 551-1124

USA, CANADA & MEXICO

Universal Sales & Marketing
230 Fifth Avenue
Suite 1212
New York, NY 10001 USA

Tel: 212-481-3500
Fax: 212-481-3534